MW01359677

THE CHEMISTRY AND MANUFACTURE OF HYDROGEN

BY

P. LITHERLAND TEED

A.R.S.M. (MINING AND METALLURGY), A.I.M.M.
MAJOR, R.A.F.

LONDON
EDWARD ARNOLD
1919

We are happy to have been able to bring such an excellent work of text back to life and to once again make it available to everyone.

ISBN:978-1-60322-038-5

www.KnowledgePublications.com

DEDICATED TO

BRIG.-GENS. E. M. MAITLAND, C.M.G., D.S.O., R.A.F.

AND

E. A. D. MASTERMAN, C.B.E., A.F.C., R.A.F.

PREFACE.

Though our national requirements are perhaps the greatest, it is noteworthy that our contribution to the technology of hydrogen is probably the least of any of the Great Powers; so, should it happen that this work in any way stimulates interest, resulting in further improvement in the technology of the subject, the author will feel himself more than amply rewarded.

The author would like to express his thanks to the Director of Airship Production for permission to publish this book, and to Major L. Rutty, R.A.F., for many helpful suggestions in the compilation of the text and assistance in correcting the proofs.

P. L. T.

Eynsford, Kent.

CONTENTS.

CHAP. PAGE

I. Hydrogen—Its Uses—Discovery, and Occurrence in Nature 1

II. The Chemical Properties of Hydrogen 9

III. The Manufacture of Hydrogen. Chemical Methods 39

IV. The Manufacture of Hydrogen. Chemico-Physical Methods 113

V. The Manufacture of Hydrogen. Physical Methods 126

Appendix. Physical Constants 145

CHAPTER I.

HYDROGEN—ITS USES—DISCOVERY, AND OCCURRENCE IN NATURE.

The Uses of Hydrogen.—The commercial production of hydrogen has received a great stimulus during the last few years owing to its being required for industrial and war purposes in quantities never previously anticipated.

The discoveries of M. Sabatier with regard to the conversion of olein and other unsaturated fats and their corresponding acids into stearin or stearic acid have created an enormous demand for hydrogen in every industrial country;[1] the synthetic production of ammonia by the Haber process has produced another industry with great hydrogen requirements, while the Great War has, through the development of the kite balloon and airship, made requirements for hydrogen in excess of the two previously mentioned industries combined.

The increase in hydrogen production has modified the older processes by which it was made, and has also led to the invention of new processes, with the result that the cost of production has decreased and will probably continue to decrease, thus allowing of its employment in yet new industries.

[1] The weight of oil hardened by means of hydrogen in Europe in 1914 probably exceeded 250,000 tons.

The Discovery of Hydrogen.—The discovery of hydrogen should be attributed to Turquet de Mayerne,[1] who in 1650 obtained, by the action of dilute sulphuric acid on iron, a gas, or "inflammable air," which we now know to have been hydrogen.

Turquet de Mayerne recognised the gas he obtained as a distinct substance. Robert Boyle[2] made some experiments with it, but many of its more important properties were not discovered until Cavendish's investigations,[3] beginning in 1766; while the actual name "Hydrogen," meaning "water former," was given to the gas by Lavoisier, who may be regarded as the first philosopher to recognise its elemental nature.

Occurrence in Nature.

Hydrogen occurs in small quantities in Nature in the uncombined state. It is found in a state of condensation in many rocks and in some specimens of meteoric iron. It is present in the gaseous discharges from oil and gas wells and volcanoes, and is also a constituent to a very minute extent of the atmosphere.

Hydrogen in the uncombined state exists in enormous masses upon the sun, and is present in the "prominences" observed in solar eclipses, while by optical means it may also be detected in many stars and nebulæ.

[1] Paracelsus, in a similar experiment in the sixteenth century, obtained the same gas, but failed to recognise it as a distinct substance.

[2] "New Experiments touching the Relation between Flame and Air," by the Hon. Robert Boyle, 1672.

[3] James Watt, the discoverer of the steam engine, did many similar experiments about the same time, but his interpretation of his results was confused by his over-elaborate theories.

In the combined state hydrogen is extremely abundant. It is present to the extent of one part in nine (by weight) in water, and is a constituent of all acids and most organic compounds.

In Rocks.—In a state of "occlusion," or molecular condensation, hydrogen is to be found in most igneous rocks in association with other gases, the total volume of occluded gases being on the average about 4·5 times the volume of the rock.

The following analyses of Sir William Tilden[1] give the composition of the occluded gases in several rocks from different parts of the world :—

Rock.	Where From.	Carbon Dioxide.	Carbon Monoxide.	Methane.	Nitrogen.	Hydrogen.
Granite	Skye	23·6	6·45	3·02	5·13	61·68
Gabbro	Lizard	5·5	2·16	2·03	1·90	88·42
Pyroxene gneiss	Ceylon	7·72	8·06	·56	1·16	12·49
Gneiss	Seringpatam	31·62	5·36	·51	·56	61·93
Basalt	Antrim	32·08	20·08	10·00	1·61	36·15

In Meteoric Iron.—An examination of certain meteoric irons, made by Sir William Ramsay and Dr. Travers,[2] showed that these contained occluded gas, and that this gas was hydrogen :—

Description of Meteorite.	Weight Taken.	Hydrogen Evolved.
Toluca	1 grm.	2·8 c.c.
Charca	,,	·28 ,,
Rancho de la Pila	,,	·57 ,,

[1] "Proc. Roy. Soc.," 1897. [2] *Ibid.*

Observing that meteoric iron contains occluded hydrogen, it is interesting to note that the examination of steel shows that it also possesses this property of condensing gases. Steel of the following composition—

	Per Cent.
Combined carbon	·810
Silicon	·080
Manganese	·050
Sulphur	·028
Phosphorus	·019
Iron (by difference)	99·013
	100·000

in pieces 6 × 1 × 1 cm. was heated (ultimate temperature 979° C.) for ten days *in vacuo* and the gases evolved analysed, with the result that they were found to have the following composition:—

	Per Cent. by Volume.
Hydrogen	52·00
Carbon monoxide	45·52
„ dioxide	1·68
Methane	·72
Nitrogen	·08
	100·00

The total weight of steel was 69·31 grammes, while the total volume of gas evolved was 19·86 c.c.[1]

An examination of a defective Admiralty bronze casting showed that there was an appreciable quantity of occluded gas in it, containing 7·6 per cent. of hydrogen by volume.[2]

[1] "Gases Occluded in Steel," by T. Baker, Iron and Steel Institute, "Carnegie Scholarship Memoirs," vol. i., 1909.

[2] "An Investigation on Unsound Castings of Admiralty Bronze," by H. C. H. Carpenter and C. F. Elam, Inst. of Metals, 1918.

In Discharge from Oil and Gas Wells.—The gas discharged from gas and oil wells contains small quantities of hydrogen, as will be seen from the following analyses of natural gas discharges in Pennsylvania, West Virginia, Ohio, Indiana, and Kansas.

AVERAGE COMPOSITION BY VOLUME.[1]

	Pa. & W. Va.	Ohio & Ind.	Kansas.
Hydrogen	·10	1·50	·00
Carbon dioxide	·05	·20	·30
Sulphuretted hydrogen	·00	·15	·00
Oxygen	trace	·15	·00
Carbon monoxide	·40	·50	1·00
Methane	80·85	93·60	93·65
Other hydrocarbons	14·00	·30	·25
Nitrogen	4·60	3·60	4·80

In Gases from Volcanoes.[2]—The nature of the gases discharged from volcanoes has been most carefully studied from about the middle of the last century, with the result that the chemical composition of the gas discharged has been determined at many different volcanoes, and at different times at the same volcano. From these investigations it would appear that in the more violent discharges there are very considerable amounts of hydrogen, while in the more placid eruptions there is little gas of any description, except steam, generally accompanied by water containing mineral salts.

[1] U.S.A. Geological Survey, "Mineral Resources of U.S.A.," 1909, 2, 297.

[2] For further information on this subject see F. W. Clarke's "The Data of Geochemistry," U.S.G.S., Bull. 616.

Below are given analyses of volcanic gas from different parts of the world by different authorities :—

From a group of fumaroles at Reykjalidh, Iceland[1] :—

Hydrogen	25·14
Oxygen	—
Nitrogen	0·72
Carbon dioxide	50·00
Sulphur dioxide	—
Sulphuretted hydrogen	24·12
	99·98

From a fumarole on Mont Pelee, Martinique[2] :—

Hydrogen	8·12
Oxygen	13·67
Nitrogen	54·94
Carbon dioxide	15·38
Sulphur dioxide	—
Carbon monoxide	1·60
Sulphuretted hydrogen	—
Methane	5·46
Argon	·71
	99·88

From Kilauea[3] :—

Hydrogen	10·2
Oxygen	—
Nitrogen	11·8
Carbon dioxide	73·9
,, monoxide	4·0
Sulphur dioxide	—
	99·9

[1] R. W. Bunsen, "Annales Chim. Phys.," 3rd ser., vol. 38, 1853.

[2] H. Moissan, "Comptes Rend.," vol. 135, 1902.

[3] A. L. Day and E. S. Shepherd, "Bull. Geol. Soc. America," vol. 24, 1913.

From Santorin[1]:—

Hydrogen	29·43
Oxygen	·32
Nitrogen	32·97
Carbon dioxide	36·42
Carbon monoxide	—
Methane	·86
Sulphuretted hydrogen	—
	100·00

In Clays.—Not only is hydrogen present in most igneous rocks, but it is to be found to a small extent in some clays. Sir William Crooks, O.M., F.R.S., was kind enough to investigate for the author the gases occluded in the celebrated "Blue Ground"—a clay in which the Kimberley diamonds are found. This clay was found to contain gas composed of 82 per cent. of carbon dioxide, the bulk of the residue being oxygen and nitrogen, with detectable traces of hydrogen.

In Air.—As is not surprising, hydrogen is present in the atmosphere to a very small extent, as will be seen from the following analysis of air under average conditions. It is doubtless derived from the sources already mentioned, and also from the decay of organic matter containing hydrogen.

The following represents the average composition of normal air:—

	Volumes per 1000.
Nitrogen	769·500
Oxygen	206·594
Aqueous vapour	14·000

[1] F. Fouque, "Santorin et ses eruptions," Paris, 1879.

	Volumes per 1000.
Argon	9·358
Carbon dioxide	·336
Hydrogen	·19
Ammonia	·008
Ozone	·0015
Nitric acid	·0005
Neon	·01
Helium	·001
Krypton	·001
Xenon	·00005

CHAPTER II.

THE CHEMICAL PROPERTIES OF HYDROGEN.

HYDROGEN in the free state has a capability of entering into combination with a large variety of substances, forming chemical compounds, while hydrogen in the combined state reacts with many other chemical compounds, forming new compounds.

Reaction of Hydrogen with Oxygen in the Free State.

By far the most important chemical reaction of hydrogen is undoubtedly that which it enters into with oxygen. When hydrogen is mixed with oxygen and the temperature of the mixed gases raised, they combine with explosive violence, producing steam. This reaction may be expressed by the following equation :—

$$2H_2 + O_2 = 2H_2O.$$

If a stream of hydrogen issues into air and a light is applied to it, it burns (in accordance with the above equation) with an almost non-luminous flame. (This reaction is, of course, reversible, i.e. a stream of air would burn in the same way in an atmosphere of hydrogen.) It was discovered by Frankland[1] that while at atmospheric pressure the flame of hydrogen burning in oxygen is almost non-luminous if the pressure is

[1] "Proc. Royal Soc.," vol. xvi., p. 419.

increased to two atmospheres the flame is strongly luminous.

The combination of oxygen and hydrogen is most violent if the two gases are present in the relative quantities given in the equation, viz. two volumes of hydrogen and one of oxygen. If one or other of the gases is in excess of these quantities the violence of the reaction is reduced and the quantity of the gas in excess of that required by the equation remains as a residue. When one gas is enormously in excess of the other a condition may arise in which the dilution is so great that on sparking the mixture no reaction takes place.[1] Mixtures of air and hydrogen in which the air is under 20 per cent. (i.e. under 4 per cent. of oxygen) of the total volume behave in this way.

This point is of importance in airships, as, providing the purity of the hydrogen in the envelope is above 80 per cent. by volume, an internal spark in the envelope will not cause an explosion, but if the quantity of hydrogen by volume falls below this amount there is a risk of explosion; hence the procedure of deflating airships when the purity has dropped to 80 per cent. hydrogen by volume.

The Temperature of Ignition of Hydrogen and Oxygen.—When the two gases are mixed in the proportion of two volumes of hydrogen and one volume of oxygen it has been found that the temperature of the mixed gases must be raised to about 580° C.[2]

[1] Schoop states that when either gas contains 6 to 8 per cent. of the other it is explosive.

[2] Victor Meyer, "Berichte," No. 16, 1893, gives the temperature of violent reaction as 612-15° C. Gautier and Helier, "Comptes Rend.," 125, 271, 1897, give about 550° C.

before explosion takes place. However, Professor Baker[1] has shown that, if the two gases are not only perfectly pure but also perfectly dry (dried by being kept in contact for as long as three weeks with anhydrous phosphoric acid) at the temperature of 1000° C., they do not combine, but even in this dry condition they will explode with an electric spark.[2] This phenomenon is of great interest, and opens a wide field of philosophic speculation, but the conditions of purity and dryness are such that this high temperature of ignition can never be attained under commercial conditions.

Professor Baker has also shown that, when a mixture of ordinary hydrogen and oxygen is exposed to the influence of strong sunlight, the two gases very slowly react, with the production of water in minute quantities.

In the experiment by which Professor Baker made this discovery he placed a mixture of these two gases in a state of great purity but not of absolute dryness (in the ratio of two volumes of hydrogen and one of oxygen) in a hard glass tube closed at one end and sealed at the other by mercury. This tube was exposed outside a south window for four months, from September to December, at the end of which time it was found, after due correction for temperature and pressure, that the mixture of the two gases had contracted by $\frac{1}{23}$ of its original volume[3] by the formation of water. A similar experiment with the gases in an exceptionally dry state,

[1] "Jour. Chem. Soc.," April, 1902.

[2] Dixon, "Jour. Chem. Soc.," vols. 97 and 98.

[3] The volume of the resulting water is almost negligible, as one volume of hydrogen and oxygen in the ratio stated produces only ·006 volume (approximately) of water.

but otherwise under exactly similar conditions, showed no such contraction.

Whether the union of hydrogen with the infiltrating oxygen of the atmosphere takes place in airship envelopes, which are comparatively transparent, has not been determined, but since in airship practice there is never more than 4 per cent. of oxygen in the envelope, it is

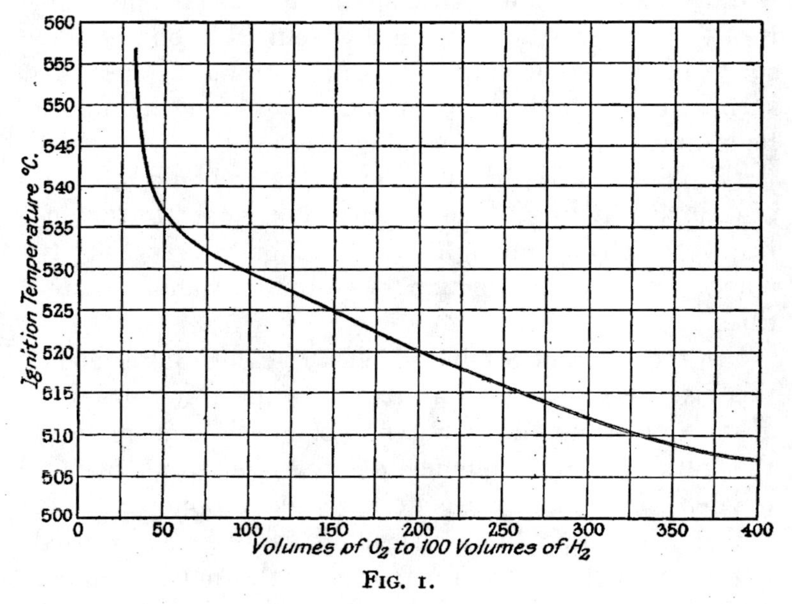

FIG. 1.

to be anticipated that such action, if it took place, would of necessity be relatively slower.

The temperature of ignition of varying mixtures of hydrogen and oxygen has been most carefully studied by Professor H. B. Dixon,[1] who, besides much very ingenious apparatus, employed the cinematograph for obtaining conclusive evidence of the conditions prevailing during explosion.

[1] "Jour. Chem. Soc.," vols. 97 and 98, and vols. 99 and 100.

By means of adiabatic compression, the temperature of ignition of different mixtures of hydrogen and oxygen was determined, with results which may be seen in Fig. 1. From a study of this curve it will be noticed that the most easily ignited mixture is not one in which the proportion of hydrogen to oxygen is as two to one, as might perhaps be expected, but when the ratio is one volume of hydrogen to four of oxygen.

IGNITION TEMPERATURES OF HYDROGEN AND OXYGEN MIXTURES.

(As determined by Prof. H. B. Dixon, M.A., F.R.S.)

(*Ignition by Adiabatic Compression.*)

Composition of Mixture. By Volume.		Ignition Temperature.
Oxygen.	Hydrogen.	° Centigrade.
33·33	100	557
40	,,	542
50	,,	536
100	,,	530
150	,,	525
200	,,	520
250	,,	516
300	,,	512
350	,,	509
400	,,	507

The temperature of ignition of a mixture fired by adiabatic compression is lower than when the same mixture is fired by being heated in a glass or silica tube at atmospheric pressure. Professor H. B. Dixon in a private communication to the author states that he found the ignition temperature of electrolytic gas under the latter conditions to be 580° C.

Besides studying the temperature of ignition of various gaseous mixtures Professor H. B. Dixon investigated the nature of explosions[1] and found that Berthelot's conception of an explosion as being an advancing locus of high pressure and of rapid chemical change, which he described as "l'onde explosive," was fundamentally correct.

Without going into detail with regard to this very interesting subject, it may be stated that "the velocity of the explosion wave in a gaseous mixture is nearly equal to the velocity of sound in *the burning gases*".

While this statement does not satisfy all cases of gaseous explosion, it may be regarded as fundamentally correct, exceptions to the rule being capable of explanation on the basis of undoubted secondary reactions.

On the basis of this relationship between the velocity of sound in the burning gases and the velocity of explosion, Professor H. B. Dixon calculated the velocity of the explosion wave in certain gaseous mixtures and also determined it experimentally, with the results given below :—

Gas Mixture.	Velocity of Explosion Wave in Metres per sec.	
	Calculated.	Found.
$8H_2 + O_2$	3554	3535
$H_2 + 3O_2$	1740	1712

While it has been said that the temperature of igni-

[1] "The Rate of Explosion in Gases," by H. B. Dixon, Bakerian Lecture, Phil. Trans. Royal Society, 1893.

tion of hydrogen and oxygen in their most readily ignited proportions must be at some point at least 500° C. in the mixture of the gases, this statement requires modification in that, though it is perfectly true in the case of a mixture of the gases contained in glass or non-porous vessels, in the presence of certain substances of a porous nature this temperature of ignition is greatly reduced. This is particularly so in the case of platinum in a spongy condition. If a piece of spongy platinum is introduced at ordinary atmospheric temperature into an explosive mixture of hydrogen and oxygen, the platinum is observed to glow and an explosion almost immediately takes place. This property is more marked if the platinum is in the spongy condition, but it is equally true if it is in the form of wire or foil.

There is no complete explanation of this phenomenon, but it has been observed that certain substances possess the property of absorbing many times their own volume of different gases, and that these absorbed gases possess a greatly increased chemical activity over their normal activity at the same temperature. Neuman and Strientz[1] found that one volume of various metals in a fine state of division is capable of absorbing the following amounts of hydrogen :—

Palladium black	502·35 volumes.
Platinum sponge	49·3 ,,
Gold	46·3 ,,
Iron	19·17 ,,
Nickel	17·57 ,,
Copper	4·5 ,,
Aluminium	2·72 ,,
Lead	·15 ,,

[1] "Zeitschrift für analytische chemie," vol. 32.

This property of certain substances, without themselves undergoing chemical change,[1] of being able to impart increased chemical activity to the gases they absorb is not confined to the metals, but is possessed by charcoal (particularly animal charcoal), magnesite brick, and probably to some extent by all porous substances. It is a subject of very great interest, and in many cases of practical importance[2] which is now becoming a subdivision of Physical Chemistry, under the name of "Surface Energy".

The Temperature Produced by the Ignition of Hydrogen and Oxygen.—In the previous paragraph the temperature at which the ignition of hydrogen and oxygen begins has been given, and now the temperature which the flame reaches will be considered.

Bunsen determined the temperature of the flame produced to be :—

Flame of hydrogen burning in air . . .	2024° C.
" " " oxygen . .	2844° C.

A later determination by Féry ("Comptes Rend.," 1902, 134, 1201) gives the values 1900° C. and 2420° C. respectively, while Bauer (*ibid.*, 1909, 148, 1756) obtained figures for hydrogen burning in oxygen varying from 2200° C. to 2300° C., according to the proportion of oxygen present.

The reason that the flame of hydrogen burning in oxygen is hotter than the flame produced in air is due

[1] It is contended by Troost and Hautefeuille that in the case of palladium the absorption of the hydrogen is chemical and not physical, palladium hydride (Pd_2H) being formed.

[2] The Bonecourt flameless boiler depends on the surface energy of magnesite brick.

to the fact that the speed of burning in oxygen is greater than in air, because of the absence of any dilution, and also because the nitrogen and other inert constituents in the air are themselves heated at the expense of the flame temperature.[1]

The calculated value for the flame temperature of hydrogen burning in air, assuming that the heat of reaction is distributed among the inert constituents of the air, is 1970° C. (Le Chatelier), and this agrees approximately with the above figures of 2024° C. and 1900° C.

A comparison between the flame temperature of hydrogen and other gases burning in air is given in the following table :—

Hydrogen[2]	1900° C.
Acetylene[3]	2548° C.
Alcohol[2]	1705° C.
Carbon Monoxide[4]	2100° C.

The Quantity of Heat Produced by Burning Hydrogen.—The temperature of ignition and the flame temperature of hydrogen have already been considered. It now only remains for the quantity of heat produced by a given weight of hydrogen to be considered in comparison with some other gases combustible in air.

[1] In the case of Zeppelin airships brought down in flames, it is not surprising that considerable amounts of molten metal have been found in the locality, observing that the melting point of aluminium is 657° C., copper 1087° C.

[2] Féry, *l.c.*

[3] Féry, *l.c.* The temperature of acetylene burning in oxygen is about 4000° C., but this arises from circumstances not present in the case of hydrogen flames.

[4] Le Chatelier.

2

1 lb. of hydrogen on combustion gives			62,100 B.T.U.[1]	
,,	marsh gas ,,	,, ,,	24,020	,,
,,	benzene ,,	,, ,,	18,090	,,
,,	carbon monoxide ,,	,,	4,380	,,

Reactions of Hydrogen with Oxygen in the Combined State.

So far the reaction of hydrogen and oxygen has only been considered when both are in the gaseous form. However, such is the attraction of hydrogen for oxygen that when the latter is in combination with some other element the hydrogen will generally combine with the oxygen, forming water and leaving the substance formerly in combination with the oxygen in a partially or wholly reduced state. Thus, oxides of such metals as iron, nickel, cobalt, tin, and lead are reduced to the metallic state by heating in an atmosphere of hydrogen.

Thus:—

$$(1)\ Fe_2O_3 + 3H_2 = 2Fe + 3H_2O$$
$$(2)\ NiO + H_2 = Ni + H_2O$$
$$(3)\ CoO + H_2 = Co + H_2O$$
$$(4)\ SnO_2 + 2H_2 = Sn + 2H_2O$$
$$(5)\ PbO + H_2 = Pb + H_2O$$

The temperature at which the reduction by the hydrogen takes place varies with the different oxides and also with the same oxide, depending on its physical condition. "Crystalline hæmatite," as the natural ferric oxide is called, requires to be at a red heat (about 500° C.) before reduction begins to take place, while if iron is precipitated from one of its salts (as ferric hydrate by

[1] The latent heat of the steam produced is included in the heat units of fuels containing hydrogen.

ammonia) the resulting ferric hydrate can be reduced to the metallic state at the temperature of boiling water.

With nickel the same variation of the temperature of reduction is noted, depending on the physical condition. Thus Moisson states that the sub-oxide of nickel (NiO) which has not been calcined, is reduced by hydrogen at 230°-240° C.; Muller, on the other hand, states that the reduction of the oxide at this temperature is not complete but only partial, but that if the temperature is raised to 270° C. a complete reduction takes place. If the oxide of nickel has been strongly heated its temperature of reduction to the metallic state is at least 420° C., in which case it is quite unsuitable for use as the catalytic agent in the hydrogenation of organic oils.

Such is the affinity of hydrogen for oxygen that hydrogen will under certain circumstances reduce hydrogen peroxide. If an acid solution of hydrogen peroxide is electrolysed, oxygen will be liberated at the positive pole (or anode), but no gas will be liberated at the negative (or cathode), for the hydrogen which is set free there immediately reduces the hydrogen peroxide in the solution to water, as shown in the following equation:—

$$H_2O_2 + H_2 = 2H_2O.$$

It has been mentioned that the temperature of reduction of the metallic oxides by hydrogen varies with the different oxides and with the physical condition of the same oxide. It might further be added that the physical condition of the hydrogen also modifies the temperature of reduction. This can be well shown by taking some artificial binoxide of tin (SnO_2) and placing it in a metal tray in a solution of slightly acidulated water. The metal tray is then connected to the negative pole of an electric supply, and another conductor placed in the liquid connected to the positive of the supply. On the current being switched on electrolysis takes place, that is to say, the water is decomposed into hydrogen and oxygen, the hydrogen being liberated on the surface of the metal tray containing the binoxide of tin, and the oxygen at the other pole. The nascent hydrogen liberated in the neighbourhood of the white tin oxide reduces it on the surface of the particle to metallic tin, in accordance with the following equation :—

$$SnO_2 + 2H_2 = Sn + 2H_2O,$$

a fact which can easily be proved by chemical means, but which is also detectable by the change of the oxide from white to the dark grey of metallic tin.

Chemical Combination of Hydrogen with Carbon.

It has been shown that if hydrogen is passed over pure carbon heated to 1150° C., direct chemical union takes place,[1] methane or marsh gas being formed :—

$$C + 2H_2 = CH_4.$$

This reaction is of some importance, as formerly in the production of blue water gas the presence of methane was entirely accounted for by the presence of hydrocarbons in the fuel. However, the experiments of Bone and Jerdan show that even if no hydrogen whatever were present in the fuel, methane would be formed if the temperature of the fuel be sufficient.

If the temperature of the carbon is somewhat hotter than 1150° C., direct union continues to take place, but the product of the reaction is not methane but acetylene.

[1] Bone and Jerdan, "Chem. Soc. Trans.," 71, 41, 1897.

Thus if a small pure carbon electric arc is made in an atmosphere of hydrogen, small quantities of acetylene are produced, but no methane.

Chemical Combination of Hydrogen with Chlorine, Bromine, and Iodine.

With Chlorine.—Hydrogen will combine with chlorine, in accordance with the following chemical equation, to make hydrochloric acid :—

$$H_2 + Cl_2 = 2HCl.$$

If the two gases are mixed in equal proportions in a diffused light and are subjected to an electric spark, the above reaction takes place with explosive violence. If a glass tube containing a mixture of the gases is heated, the same reaction takes place with violence.

If a mixture of hydrogen and chlorine at atmospheric temperature is exposed to strong sunlight, hydrochloric acid is immediately formed, with the characteristic explosion. Investigation of this increase in the chemical activity of hydrogen and chlorine in the presence of sunlight has shown that it is the actinic rays which produce the phenomenon ; thus if the rays which are present at the blue and violet end of the spectrum are prevented from reaching the mixture of the gases by protecting this by a red glass screen, no reaction between them takes place. When sunlight is not available, the explosive combination of these two gases can be shown by exposing a mixture of them in a glass vessel to the light of burning magnesium, such as is frequently used by photographers.

The remarks which have already been made with regard to the reduction in chemical activity of hydrogen and oxygen when perfectly dry apply also in the case of hydrogen and chlorine.

While referring to the production of chemical union between hydrogen and chlorine brought about by the influence of light, attention may be drawn to what is known as the "Draper Effect," which is best demonstrated in the following apparatus:—

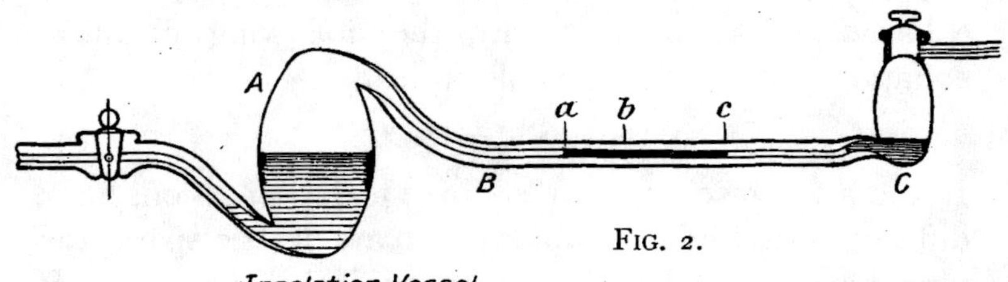

FIG. 2.

The mixed gases, in the ratio of one volume of hydrogen to one of chlorine, are contained in a flat glass bulb A, called the insolation vessel. The lower part of the insolation vessel usually contains some water saturated with the two gases. The capillary tube BC contains a thread of liquid *ac*, to serve as an index. Under the influence of a flash of light the thread of liquid *ac* is pushed outwards, to return immediately to its original position. Thus, *a* travels to *b*, and immediately returns to *a*. With every flash of light the same phenomenon takes place. At the time of its discovery (1843, "Phil. Mag.," 1843, iii., 23, 403, 415) the reason for this sudden rise in pressure was not understood, but careful investigation by J. W. Mellor and W. R. Anderson[1] has shown that at each flash minute quantities of hydrochloric acid are formed, with the production of a little heat, thus causing a rise in pressure until it is dispersed

[1] "Jour. Chem. Soc.," April, 1902.

—in fact, the Draper effect may be likened to a very small explosion without sufficient energy to propagate itself throughout the gas.

Such is the attraction of chlorine for hydrogen that even when the latter is in combination with some other element the chlorine often will combine with the hydrogen, liberating that element. Thus, if chlorine is passed through turpentine, the carbon is liberated, in accordance with the following equation :—

$$C_{10}H_{16} + 8Cl_2 = 10C + 16HCl.$$

Again, at ordinary temperatures and in ordinary diffused light, but more rapidly in sunlight or other light of actinic value, chlorine will decompose water, liberating oxygen, in accordance with the following equation :—

$$2H_2O + 2Cl_2 = 4HCl + O_2.$$

The combination of hydrogen with chlorine is attended with the evolution of heat. According to Thomsen, the combination of 1 gramme of hydrogen with 35·5 grammes of chlorine is attended with the evolution of 22,000 gramme-calories of heat.

With Bromine.—The element bromine will combine with hydrogen to form hydrobromic acid, in accordance with the following equation :—

$$H_2 + Br_2 = 2HBr.$$

This reaction between hydrogen and bromine is in many respects comparable with the combination of hydrogen with chlorine, but unlike the latter, the reaction cannot be brought about by sunlight. However, if the two gases are heated, they will combine, but their combination is attended with the evolution of less heat than in the case of chlorine. Thomsen states that the combination of 1 gramme of hydrogen with 80 grammes of bromine (liquid) is attended with the evolution of 8440 gramme-calories of heat.

With Iodine.—Hydrogen will combine with iodine, in accordance with the following equation, providing the iodine is in the form of vapour and the mixture of the two gases is strongly heated in the presence of spongy platinum :—

$$H_2 + I_2 = 2HI.$$

Thomsen has shown that this combination, unlike the two previous ones, is not attended with evolution of heat, but by the absorption of it. Thus when 1 gramme of hydrogen combines with 127 grammes of iodine (solid), 6040 gramme-calories of heat are absorbed.

Chemical Combination of Hydrogen with Sulphur, Selenium, and Tellurium.

With Sulphur.—If a mixture of sulphur vapour and hydrogen is passed through a tube heated to at least 250° C., a chemical union of the two elements takes place, in accordance with the equation—

$$H_2 + S = H_2S.$$

The resulting gas, which is known as "sulphuretted hydrogen," has a characteristic and extremely unpleasant odour, and is poisonous when inhaled. According to Thénard, respiration in an atmosphere containing $\frac{1}{800}$ part of its volume of sulphuretted hydrogen is fatal to a dog, and smaller animals die when half that quantity is present.

Sulphuretted hydrogen is an inflammable gas, and will burn in air, in accordance with the following equation :—

$$2H_2S + 3O_2 = 2SO_2 + 2H_2O,$$

producing sulphur dioxide and water.

If the gas is mixed with oxygen in the proportions required by the equation, and subjected to an electric spark, it explodes with violence, giving the same products as when burnt in air.

Sulphuretted hydrogen is soluble in water at 0° C. to the extent of 4·3706 parts by volume per unit volume of water.

The density of sulphuretted hydrogen is 17 times that of hydrogen.

With Selenium.—When selenium is heated to 250° C. with hydrogen, chemical union results, with the production of selenuretted hydrogen :—

$$H_2 + Se = H_2Se.$$

The resulting gas is colourless, resembling sulphuretted hydrogen in smell and in its chemical properties. It is, however, much more poisonous than the former gas.

Selenuretted hydrogen is inflammable and burns in the same way as sulphuretted hydrogen. If the gas is strongly heated it breaks up into its two constituents, the selenium being deposited in the crystalline form.

Selenuretted hydrogen is soluble in water at 13·2° C. to the extent of 3·31 parts by volume per unit volume of water.

The density of selenuretted hydrogen is 40·5 times that of hydrogen.

With Tellurium.—When tellurium is heated to 400° C. in hydrogen, the elements combine, forming hydrogen telluride :—

$$H_2 + Te = H_2Te.$$

This gas, like sulphuretted and selenuretted hydrogen, is both offensive smelling and poisonous. Like selenuretted hydrogen, on strongly heating it is decomposed into its components, the tellurium being deposited in the crystalline form.

Telluretted hydrogen is soluble in water to some extent, but in course of time the telluretted hydrogen is decomposed and tellurium deposited.

The density of telluretted hydrogen is 63·5 times that of hydrogen.

Chemical Combination of Hydrogen with Nitrogen, Phosphorus, and Arsenic.

With Nitrogen.—Donkin has shown that when a mixture of hydrogen and nitrogen is subjected to the silent electric discharge, a partial union of the two gases takes place, with the formation of ammonia :—

$$N_2 + 3H_2 = 2NH_3.$$

However, this reaction could in no way be regarded as commercial, as the quantity of ammonia produced after the gases have long been subjected to the silent electric discharge is only just sufficient to be identified by the most delicate means.

Recent investigations have, however, shown that if the two gases are mixed and subjected to very great pressure (1800 lb. per sq. inch) in the presence of a catalytic agent, union to an appreciable extent takes place. This process, which is now being used on a commercial scale in Germany, is known as the Haber process, but few details as to the method of operation are available. In the earlier stages of the working of this process the catalytic agent was probably osmium, but it is considered doubtful if this is still being employed.

THE USES OF AMMONIA.

Such is the importance of ammonia in the existence of a modern country that it is desirable that some account of its use should be given, observing that it is not improbable that the Haber process may be put into operation in this country in the near future, consequently enormously increasing the demand for the commercial production of hydrogen.

Ammonia or its salts are employed in a variety of ways in many trades. From it nitric acid, the vital necessity for the manufacture of all high explosives, can be made; it is an essential for the Brunner Mond or Solvay ammonia soda process for the production of alkali; in the liquid form it is employed all over the world in refrigerating machinery, but its enormous and increasing use is in agriculture, where, in the form of sulphate of ammonia, it constitutes one of, if not the most important chemical manures known to man. During the year 1916 350,000 tons of ammonium sulphate were produced in this country, the larger proportion of which was consumed in agriculture—a proportion likely to increase and not diminish if the demand for home production of food continues.

PROPERTIES OF AMMONIA.

Ammonia is a strongly smelling gas, possessing a most characteristic odour. It is lighter than air; taking the density of hydrogen as 1, air is 14·39, and ammonia 8·5. Ammonia is not in the ordinary sense combustible in air, but if the air is heated or oxygen is supplied it will burn with a feeble, almost non-luminous flame, in accordance with the following equation :—

$$4NH_3 + 3O_2 = 2N_2 + 6H_2O.$$

Ammonia is strongly basic, i.e. it possesses the property of combining with acids to make neutral salts. Thus with the common acids—sulphuric acid, hydrochloric acid and nitric acid—it forms salts, in accordance with the following equations :—

$$2NH_3 + H_2SO_4 = (NH_4)_2SO_4,$$
$$NH_3 + HCl = (NH_4)Cl,$$
$$NH_3 + HNO_3 = (NH_4)NO_3.$$

Among the physical properties of ammonia the outstanding features are its solubility in water, its absorption by charcoal, and its liquefaction.

Solubility of Ammonia in Water. — Ammonia is very soluble in water. Its solubility decreases with increase of temperature, and, as is of course natural, increases with increase of pressure. The following table for the solubility of ammonia in water is interesting :—

Temperature.	Grammes of NH_3 Dissolved in 1 c.c. of Water.	C.c. of NH_3 at 0° C. and 760 mm.
0° C.	·875	1148
8	·713	923
16	·582	764
30	·403	529
50	·229	306

A feature of the absorption of ammonia by water is the reduction of the specific gravity of the solution. Thus at 15° C. a saturated solution containing 34·95 per cent. of the gas by weight has a density of ·882, while pure water at the same temperature has a density of ·99909.

Absorption of Ammonia by Charcoal.—Reference to the surface energy of charcoal has already been made. Its absorption of ammonia is very considerable, but varies with the physical condition of the charcoal, as well as with the material from which it has been made. Saussure found that freshly ignited boxwood absorbs about 90 times its own volume of ammonia, while Hunter has shown that freshly prepared charcoal made from cocoanut shell absorbs about 171 times its own volume of ammonia.

Liquefaction of Ammonia.—Ammonia is an easily liquefiable gas, and consequently it is owing to this property that it is employed in refrigerating plants on land and in ships, for by the rapid evaporation of the liquid gas a high degree of cold may be obtained. The critical temperature of ammonia, i.e. that temperature above which by mere pressure it cannot be liquefied, is 131° C. At this temperature a pressure of approximately 1700 lb. per sq. inch must be applied to produce liquefaction ; if, however, the temperature is below the critical one for the gas, the pressure required for liquefaction is greatly reduced. Thus, if the ammonia is cooled to 15·5° C., a pressure of 101 lb. per sq. inch is required, while if the gas is cooled to 0° C., a pressure of only 61·8 lb. per sq. inch will effect liquefaction. Liquid ammonia is a colourless, mobile liquid. It boils at − 33·7° C., and at 0° C. has a specific gravity of 0·6234.

At -75° C. liquid ammonia solidifies into a white crystalline solid.

With Phosphorus.—If red phosphorus is gently heated in a stream of hydrogen, direct chemical union takes place to a small extent, with the production of a gas termed "Phosphoretted Hydrogen" or "Phosphine":—

$$2P + 3H_2 = 2PH_3.$$

Phosphine is an offensive smelling, poisonous gas which in the pure state is not spontaneously inflammable. However, its temperature of ignition is very low; thus, if a stream of phosphine is allowed to impinge in air on a glass vessel containing boiling water, it will immediately burst into flame, burning with considerable luminosity, in accordance with the equation:—

$$PH_3 + 2O_2 = HPO_3 + H_2O.$$

Phosphine possesses an exceedingly interesting reaction with oxygen. Thus, if a mixture of phosphine and oxygen is subjected to a sudden reduction in pressure at ordinary atmospheric temperature, chemical combination immediately takes place with explosive violence, in accordance with the equation already given.

Phosphine, which is produced in small quantities in the Silicol process for making hydrogen,[1] has under certain conditions a deteriorating effect on cotton fabrics, not as an immediate action but as a secondary reaction. The examination of a balloon envelope which burst at Milan[2] in 1906 showed that at some spots the material could be easily torn, while over the greater portion it

[1] The total volume of phosphine and arsine does not exceed ·025 per cent. and is usually about ·01 per cent.

[2] Namias, "L'Ind. Chim.," 1907, 7, 257-258; "Chem. Cent.," 1907, 2, 1460-1461.

showed a great resistance to tearing. The damaged spots were found to be impregnated with phosphoric acid and arsenic acid, produced by the oxidation of the phosphine and arsine contained in the hydrogen with which the balloon had been inflated.

Phosphine in small quantities in hydrogen containing over 1 per cent. of oxygen attacks copper, producing an acid liquid which has a most corrosive action on fabric. However, it does not appear under these circumstances to have any action on aluminium or zinc; consequently any metal parts inside the envelope of an airship should be of aluminium. Phosphine under the above conditions attacks hemp and other textiles which have been treated with copper compounds, but it does not appear to have any action on fabrics free from copper compounds or copper or brass fastenings.

Though it has been stated that phosphine is not spontaneously inflammable, with quite small admixtures of liquid hydrogen phosphide it immediately bursts into flame on coming into contact with air.

Phosphine produced by the reaction of water on calcium phosphide always contains a quantity of the liquid hydrogen phosphide sufficient to make the gas spontaneously inflammable. Use of this property is made in the Holmes' Light used at sea as a distress signal, and also as a marker at torpedo practice.

Phosphine is soluble in water to a slight extent. The solution of phosphine in water is not very stable, particularly in strong light, when it breaks up, depositing red phosphorus.

The density of phosphine is 17·5 times that of hydrogen.

With Arsenic.—Hydrogen does not directly combine with arsenic, but if an arsenic compound is in solution in a liquid in which hydrogen is being generated, i.e. hydrogen in the nascent state, chemical union takes place. Thus, if arsenious oxide is dissolved in dilute hydrochloric acid and a piece of metallic zinc is added, the hydrogen produced by the action of the acid on the zinc will combine with the arsenic, in accordance with the following equation :—

$$As_4O_6 + 12H_2 = 4AsH_3 + 6H_2O.$$

The gas produced, which is called "Arsine" or "Arsenuretted Hydrogen," is unpleasant smelling and poisonous. It burns in air with a lilac-coloured but not very luminous flame, thus :—

$$4AsH_3 + 6O_2 = As_4O_6 + 6H_2O.$$

If the gas is strongly heated it is decomposed and elemental arsenic deposited.

Arsine is produced to a small extent in the Silicol process of making hydrogen, and has a deteriorating effect on fabric (see phosphine), while with many metals it is decomposed, arsenic being deposited and hydrogen liberated. It can be liquefied easily (the liquid gas boiling at $-54 \cdot 8°$ C), and it solidifies at $-113 \cdot 5°$ C. Arsine is soluble in water, one volume of water at 0° C. dissolving 5 volumes of arsine. The density of arsine is 39 times that of hydrogen.

Chemical Combination of Hydrogen with Lithium, Sodium, Potassium, Magnesium, Calcium, and Cerium.

The chemical combination of hydrogen has so far only been considered with regard to a few non-metallic elements, but now a new series of reactions will be considered in which hydrogen combines chemically with a metal. These metals are those of the alkaline and alkaline earth group.

With Lithium.—If hydrogen is passed over metallic lithium at about 200° C., the hydrogen is absorbed, not as hydrogen is absorbed by platinum, etc., but chemically absorbed, in accordance with the following equation:—

$$4Li + H_2 = Li_4H_2.$$

If the resulting lithium hydride is allowed to cool and is placed in water it becomes a source of hydrogen, not only giving up what it has already received, but also a volume twice as much as this, which it has derived from the water, as may be seen in the following equation:—

$$Li_4H_2 + 4H_2O = 4LiOH + 3H_2.$$

With Sodium.—Under similar circumstances the metal sodium absorbs hydrogen with the production of a hydride:—

$$4Na + H_2 = Na_4H_2.$$

This hydride, like that of lithium, behaves in a similar manner with water. It, however, has another interesting property in that if sodium hydride is heated *in vacuo* to about 300° C., the whole of the hydrogen is given off and metallic sodium again remains.

With Potassium.—If the metal potassium is heated in the presence of hydrogen, a hydride is formed:—

$$4K + H_2 = K_4H_2.$$

This hydride has the same characteristic reaction with

water, but it has a distinctive reaction, in that on exposure to air it catches fire :—

$$2K_4H_2 + 9O_2 = 4K_2O_4 + 2H_2O.$$

With Magnesium.—If hydrogen is passed over hot metallic magnesium the hydrogen is absorbed :—

$$Mg + H_2 = MgH_2.$$

This hydride is decomposed with water, with the production of magnesium hydrate and hydrogen :—

$$MgH_2 + 2H_2O = Mg(OH)_2 + 2H_2.$$

With Calcium.—If hydrogen is passed over hot metallic calcium the hydrogen is absorbed :—

$$Ca + H_2 = CaH_2.$$

The hydride is decomposed by water, according to the equation—

$$CaH_2 + 2H_2O = Ca(OH)_2 + 2H_2.$$

Calcium hydride, unlike the metallic hydrides already mentioned, is a commercial possibility, and under the name of "Hydrolith" has been used by the French Army in the field for the inflation of observation balloons. Its use for this purpose is governed by French patent No. 327878, 1902, in the name of Jaubert.

With Cerium.—If hydrogen is passed over hot metallic cerium the hydrogen is absorbed :—

$$Ce + H_2 = CeH_2.$$

This hydride is decomposed with water in the same manner as calcium hydride, but as a source of hydrogen it is far too rare to be employed.

However, if an alloy of cerium with magnesium and

aluminium is heated below its melting point in a stream of hydrogen, the latter is absorbed, with the formation of cerium hydride within the alloy, which, after cooling, possesses to a remarkable degree the property of emitting sparks when rubbed with any rough surface. These sparks are sufficiently hot to ignite coal gas and petrol vapour, hence the employment of this hydrogenated alloy in the patent lighters which have of recent years become so common in this country.

Chemical Combination of Hydrogen with Animal and Vegetable Oil.

Owing to the discoveries of M. Sabatier a new use has been found for hydrogen, and a vast and ever-growing industry created, known as " fat hardening ".

The chief uses for animal and vegetable fats are for the making of candles, soap, and edible fats such as are incorporated in butter substitutes, sold generically under the name of " Margarine ".

Animal and vegetable fats are generally mixtures of a certain number of complicated organic chemical compounds, amongst the chief of which may be mentioned linolein, olein, stearin, and palmitin. The physical properties of these compounds are somewhat different. Thus, those containing considerable proportions of stearin and palmitin are usually solid at atmospheric temperature, while those in which the chief constituent is either linolein or olein are liquids at such temperature.

These chemical compounds—linolein, olein, stearin, and palmitin—are what are known as "glycerides," i.e. they are compounds of glycerine with an organic acid.

Now since glycerine is of great value in a variety of ways, chiefly for the production of nitro-glycerine, it is customary to split these glycerides up into glycerine and their organic acid before indulging in any other process. This may be accomplished by the use of superheated steam. Thus, when such steam is blown through palmitin the following reaction takes place :—

$$\underset{\text{Palmitin}}{C_3H_5(C_{16}H_{31}O_2)_3} + \underset{\text{Steam}}{3H_2O} = \underset{\text{Palmitic acid}}{3H(C_{16}H_{31}O_2)} + \underset{\text{Glycerine}}{C_3H_5(HO)_3}.$$

Or through olein :—

$$\underset{\text{Olein}}{C_3H_5(C_{18}H_{33}O_2)_3} + \underset{\text{Steam}}{3H_2O} = \underset{\text{Oleic acid}}{3H(C_{18}H_{33}O_2)} + \underset{\text{Glycerine}}{C_3H_5(HO)_3}.$$

The physical properties of these organic acids are very interesting and important. Their melting points are :—

Palmitic acid	Melting point, 62·6° C.
Stearic acid	,, ,, 69·3° C.
Oleic acid	,, ,, 14·0° C.

Now this oleic acid, owing to its low melting point, is not of great value, as it cannot be used for candles. However, the discoveries of M. Sabatier have shown that under certain conditions of temperature and in the presence of nickel or cobalt (which themselves undergo no permanent change), the low melting linoleic and oleic acids may be converted into stearic acid by the introduction of hydrogen into the liquid organic acid. Thus :—

$$\underset{\text{Oleic acid}}{C_{17}H_{33}COOH} + H_2 = \underset{\text{Stearic acid}}{C_{17}H_{35}COOH}.$$

The nickel in this process may be introduced into the liquid organic acid by merely adding spongy nickel to the molten oleic acid ; or as a volatile compound known as "Nickel Carbonyl" it may be blown in together with the hydrogen.

In either case, for the conversion of the linoleic and oleic acids into stearic acid, the temperature of the acids should be between 200° and 220° C. When the nickel is introduced, in the form of carbonyl, at the same time as the hydrogen, the carbonyl is decomposed into metallic nickel and carbon monoxide—the latter taking no part whatever in the reaction and being available for the production of further nickel carbonyl.

The nickel which is used in this process performs merely a catalytic function and does not of itself undergo permanent change. However, its catalytic property may be destroyed either by the method by which it is prepared or by certain impurities in the hydrogen with which the hydrogenation is carried out. While it is not important that the hydrogen should be very pure—in fact, it may contain carbon monoxide, nitrogen, carbon dioxide, and methane—it is absolutely essential that it should be entirely free from sulphur dioxide, sulphuretted hydrogen, and other sulphur compounds, bromine, chlorine, iodine, hydrochloric acid, arsenuretted hydrogen, selenuretted hydrogen, and teluretted hydrogen.

If the nickel is introduced into the fatty acid in the solid form it is important that it should be absolutely free from sulphur, selenium, tellurium, arsenic, chlorine, iodine, bromine. Further, it is important that the nickel should have been prepared by the reduction of the oxide at a temperature not exceeding 300° C., and should not have been long exposed to the air prior to its use.

The weight of nickel used is about 0·1 part to 100 parts of oil or fatty acid; however, larger quantities do no harm. After the hydrogenation of the fatty acid or oil, practically the whole of the nickel is recovered by merely filtering the hot oil or fatty acid.

In this note the use of hydrogen in the fat hardening industry has been described with particular reference to the conversion of the unsaturated oleic and linoleic fatty acids into stearic acid. However, what has been said in regard to this matter is equally applicable to the conversion of olein and linolein into stearin, cotton-seed and most fish oils being quite easily converted into solid fats.

CHAPTER III.

THE MANUFACTURE OF HYDROGEN. CHEMICAL METHODS.

THE PRODUCTION OF HYDROGEN.

WHILE all the processes described yield hydrogen, some are of merely laboratory use, others of commercial use, and yet others of use for the generation of hydrogen for war purposes, under conditions where rapidity of production and low weight of reagents are more important than the cost of the final product.

Where hydrogen is wanted for commercial purposes two types of process will generally be found most useful: the electrolytic, where not more than 1000 cubic feet of hydrogen are required per hour and conditions are such that the oxygen produced can be either advantageously used or sold locally; the Iron Contact process, the Linde-Frank-Caro process, or the Badische Anilin Catalytic process, where yields of 3000 and more cubic feet are required per hour. However, local conditions and the requirements of a particular trade may make some of the other processes the more desirable.

For war hydrogen may be economically produced at a base, and used there for the inflation of airships, or the filling of high-pressure bottles for transport to the Kite Balloon Sections in the field. Where transport conditions are difficult it may be advantageous to generate the hydrogen on the field at the place where it will be used; then, probably, the Silicol, Hydrogenite or Hydrolith processes will have the advantage, but here again it is not possible to speak with any great precision, as local conditions, even in war, must have great influence on the selection of the most suitable process.

The production of hydrogen can be accomplished by a large variety of methods, which may be divided into two main classes, viz. chemical and physical, while there is an intermediate class in which the production of hydrogen is accomplished in two stages, one being chemical and the other physical.

Chemical Methods of Producing Hydrogen.

The chemical methods of producing hydrogen may be divided into four classes:—

1. Methods using an acid.
2. Methods using an alkali.
3. Methods in which the hydrogen is derived from water.
4. Methods in which the hydrogen is produced by methods other than the above.

(1) Methods Using an Acid.

With Iron.—If dilute sulphuric acid is brought into contact with iron, chemical action takes place, with the production of hydrogen and ferrous sulphate, in accordance with the following equation:—

$$Fe + H_2SO_4 = H_2 + FeSO_4.$$

Theoretically, to produce 1000 cubic feet of hydrogen at 30 inches barometric pressure and 40° F. by this process, 155 lb. of iron and 272 lb. of pure sulphuric acid are required, or a total weight of pure reagents equal to 427 lb. per 1000 cubic feet of hydrogen produced. From the figures given above, the approximate cost of material per 1000 cubic feet of hydrogen can be calculated if the prevailing prices of iron and sulphuric acid are known. Of course, pure sulphuric acid is not an essential for the process, but allowance for the impurity of the sulphuric acid and iron must be made in any calculation for cost or weight.

The hydrogen produced by this method varies considerably in purity. It is liable to contain methane to an extent which depends on the carbon content of the iron; it may also contain phosphine, depending on the phosphorus content of the iron, sulphuretted hydrogen, depending on the sulphur content of the iron, and traces of silicon hydride, depending on the silicon content of the iron. It is also liable to contain arsine or arsenuretted hydrogen, depending on the arsenic content of the sulphuric acid, the commercial acid frequently containing considerable amounts of this impurity. Unless specially treated, the hydrogen produced is always acid, and therefore unsuitable for balloon and airship purposes.

The impure gas produced by this method may be purified by being passed through or scrubbed by water; this will remove much of the acid carried by the gas, dust, and some of the methane, phosphine, arsine, and sulphuretted hydrogen. If after this treatment the gas is passed through a solution of a lead salt, the remaining acidity and sulphuretted hydrogen can be removed. This method of the treatment of the impure gas is covered by English patent 16277, 1896, in the names of Pratis and Marengo. Further patents in connection with this method of producing hydrogen have been taken out by Williams (English patent 8895, 1886), Hawkins (English patent 15379, 1891), Pratis and Marengo (English patent 15509, 1897), Hawkins (English patent 25084, 1897), and Fielding (English patent 17516, 1898).

With Zinc.—If dilute sulphuric acid is brought into contact with zinc, chemical action takes place, with the production of zinc sulphate and hydrogen, in accordance with the following equation :—

$$Zn + H_2SO_4 = H_2 + ZnSO_4.$$

Theoretically, to produce 1000 cubic feet of hydrogen at 30 inches barometric pressure and 40° F. by this process, 180 lb. of zinc and 272 lb. of pure sulphuric acid are required, or a total weight of pure reagents equal to 452 lb. per 1000 cubic feet of hydrogen produced.

The hydrogen produced by this process is liable to fewer impurities than when iron is used, but it is always acid and liable to contain arsine if commercial sulphuric acid is used.

The zinc sulphate produced in this process can be turned more easily to commercial account than iron sulphate. If to the solution of the zinc sulphate resulting from the process sodium carbonate or sodium hydrogen carbonate is added, a precipitate of hydrated zinc basic carbonate or zinc carbonate is obtained, which on ignition in a furnace yields zinc oxide (commercially known as "zinc white"), water, and carbon dioxide. Zinc white has a commercial value as a basis or body in paints ; it has one great advantage over white lead, which is used for the same purpose, in that it is far less poisonous. This method of treatment of the residual zinc sulphate is the subject of a patent by Barton (English patent 28534, 1910).

The previous list of patents for the reaction of iron and sulphuric acid also cover the use of zinc and sulphuric acid for the production of hydrogen.

There are other metals which will yield hydrogen with sulphuric acid, such as cadmium and nickel, while many metals will yield hydrogen with hydrochloric acid, such as tin, nickel, and aluminium. However, these reactions cannot be regarded as commercial means of producing hydrogen.

(2) Methods Using an Alkali.

With Zinc.—If a solution of caustic soda in water is brought into contact with metallic zinc, chemical reaction takes place, with the production of sodium zincate and hydrogen. The reaction is expressed in the following equation :—

$$Zn + 2NaOH = H_2 + Na_2ZnO_2.$$

Theoretically, to produce 1000 cubic feet of hydrogen at 30 inches barometric pressure and 40° F., 180 lb. of zinc and 224 lb. of pure caustic soda are required, or a total weight of pure reagents equal to 404 lb. per 1000 cubic feet of hydrogen produced.

The hydrogen produced by this process is generally very pure, but, depending on the purity of the zinc, it is liable to contain arsine. As the gas is alkaline, owing to the caustic soda carried in suspension, it requires to be scrubbed to make it suitable for balloons and airships.

A modification of this process has been the subject of a patent. Zinc as a fine powder is mixed with dry slaked lime; then when hydrogen is required, the mixture is heated in a retort and hydrogen is evolved, the reaction being expressed :—

$$Zn + Ca(OH)_2 = H_2 + CaZnO_2.$$

In this modification of the process to produce 1000 cubic feet of hydrogen at 30 inches barometric pressure and 40° F., 180 lb. of zinc and 207 lb. of slaked lime are required, or a total weight of pure reagents equal to 387 lb. per 1000 cubic feet of hydrogen produced.

By the substitution of magnesium hydroxide instead of slaked lime a similar reaction takes place, but the total weight per 1000 cubic feet of hydrogen produced is reduced to 341 lb.

This process, with its modification, is covered by a patent by Majert and Richter (English patent 4881, 1887), and is primarily intended as a process for the generation of hydrogen in the field for the inflation of observation balloons.

THE HYDRIK OR ALUMINAL PROCESS.

With Aluminium.—If a solution of caustic soda is brought into contact with metallic aluminium, chemical reaction takes place, with the production of sodium aluminate and hydrogen, in accordance with the following equation :—

$$2Al + 6NaOH = 3H_2 + 2Al(ONa)_3.$$

Theoretically, to produce 1000 cubic feet of hydrogen at 30 inches barometric pressure and 40° F., 50 lb. of aluminium and 225 lb. of pure caustic soda are required, or a total weight of pure reagents equal to 275 lb. per 1000 cubic feet of hydrogen.

The hydrogen produced by this process is generally

very pure, but the gas is frequently alkaline from minute traces of caustic soda carried in suspension, which must be removed by scrubbing with water before the hydrogen is suitable for balloons and airships.

THE SILICOL PROCESS.

With Silicon.—If a solution of caustic soda is brought into contact with elemental silicon, chemical reaction takes place, with the production of sodium silicate and hydrogen. The following equation was supposed to represent the reaction :—

$$Si + 2NaOH + H_2O = Na_2SiO_3 + 2H_2.$$

Theoretically, to produce 1000 cubic feet of hydrogen at 30 inches barometric pressure and 40° F., 38·8 lb. of silicon and 111 lb. of pure caustic soda are required, or a total weight of pure reagents equal to 149·8 lb. per 1000 cubic feet of hydrogen.

The gas produced by this process is singularly pure, generally containing 99·9 per cent. hydrogen by volume (if the water vapour is removed before analysis), ·01 per cent. of arsine and phosphine, ·005 per cent. acetylene, the remaining impurity being air, which is introduced in the powdered silicon and also in solution in the water.

In working this process practically, pure silicon is not used, high-grade ferro-silicon, containing 82-92 per cent. silicon, being employed. As will be seen from the above equation, theoretically 2·86 parts of anhydrous caustic soda by weight should be used for one part of silicon. However, in working in practice, one part of pure silicon and 1·7 parts of pure caustic soda are employed. This discrepancy between the theoretical quantity of soda and that actually used has been investigated by the author, who originally considered that the following reaction might be taking place :—

$$Si + 2H_2O = SiO_2 + 2H_2.$$

That is to say, the silicon was being oxidised by the oxygen of the water, and hydrogen liberated.

The first experiment performed was the heating of the ferro-silicon[1] (92 per cent. Si) in a flask with boiling water; the resulting steam was condensed, but there was no residual gas. Therefore it was concluded that at the temperature of boiling water no reaction between ferro-silicon and water took place.

Remembering that the temperature of the caustic soda solution used in the silicol process is above 100° C., frequently rising to 120° C., it was thought that a higher temperature might perhaps produce the suspected reaction; ferro-silicon was accordingly heated in an atmosphere of steam in an electric resistance furnace to a temperature of 300° C., but still no hydrogen was produced. Consequently it was concluded that the explanation of the smaller consumption of caustic soda than would be anticipated from theoretical considerations must be explained on some basis other than the reaction of silicon with water.

The next experiment attempted was the heating of ferro-silicon with sodium silicate, i.e. with a pure form of the product of the usual equation. When ferro-silicon was heated with an aqueous solution of pure sodium mono-silicate, considerable quantities of hydrogen were

[1] The ferro-silicon employed was of French manufacture. I have since found that some high-grade Canadian ferro-silicons give traces of hydrogen with water under the conditions cited in the experiments.

evolved, thus warranting the conclusion that the ordinary equation—

$$Si + 2NaOH + H_2O = Na_2SiO_3 + 2H_2$$

is not entirely correct, and that a silicate richer in silica than that indicated in the equation was formed, and that probably the following reaction proceeds to some extent :—

$$Si + Na_2SiO_3 + 2H_2O = Na_2Si_2O_5 + 2H_2.$$

Assuming this second reaction to take place at the same time as the first, the reaction can be expressed :—

$$2Si + 2NaOH + 3H_2O = Na_2Si_2O_5 + 4H_2,$$

which is equivalent to 1000 cubic feet of hydrogen at 30 inches barometric pressure and 40° F. being produced by 38·8 lb. of silicon and 55·5 lb. of caustic soda, the ratio of pure caustic soda to pure silicon being as 1·43 is to 1.

Using a plant producing about 30,000 cubic feet of hydrogen per hour, it was found that 1·9 parts of caustic soda (76 per cent. NaOH) to 1 part of Canadian ferro-silicon (84 per cent. Si) gave very satisfactory results, the ratio of the pure reagents being as 1·72 parts of caustic soda by weight to 1 part of silicon.

Theoretically, 22·5 cubic feet of hydrogen should have been produced per lb. of the commercial ferro-silicon used, but in practice it was found that 20·7 cubic feet were obtained, the discrepancy of 1·8 cubic feet being to some extent accounted for by the protective action of impurities, loss through leaks and also by hydrogen being mechanically carried away by the water used for cooling the issuing hydrogen.

Description of Silicol Plant.—The essentials of a silicol plant are shown in the diagram (Fig. 3). The requisite quantity of caustic soda is placed in the tank on the right and the necessary water added to it to make a 25 per cent. solution. To assist solution there is a

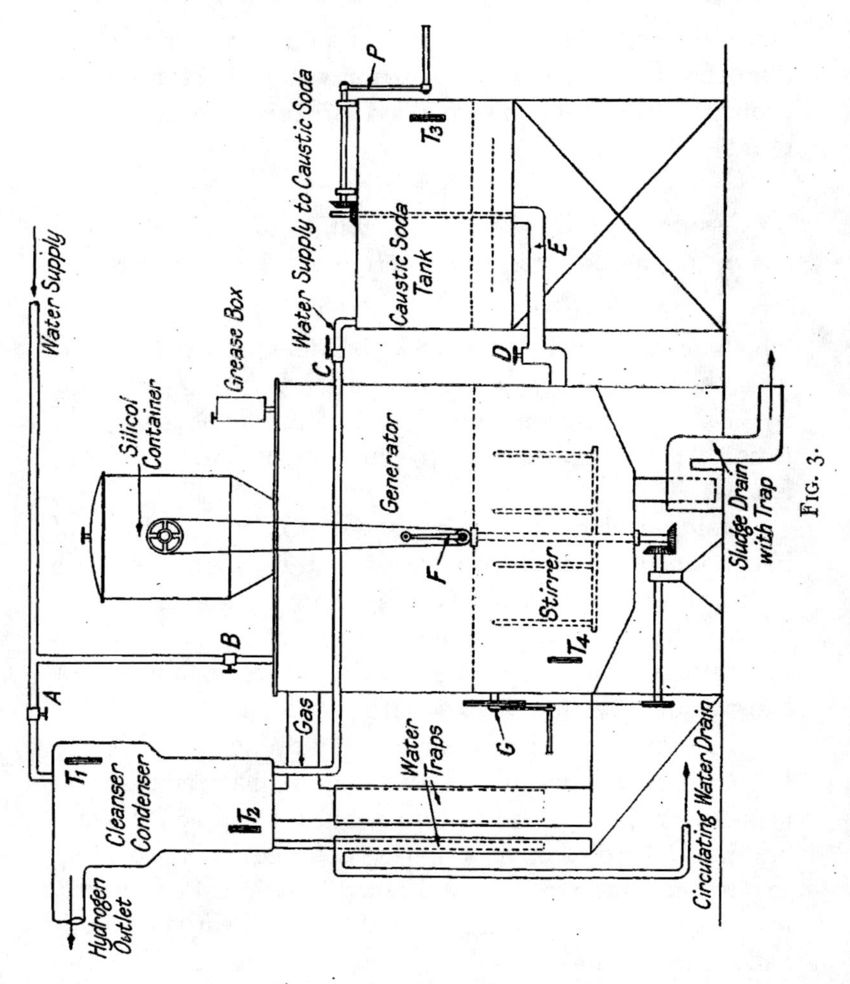

FIG. 3.

stirrer in this tank, which, in small plants, is hand-operated and in large ones power-operated. When the whole of the caustic soda has gone into solution, which it readily does as a result of the heat of solution and the stirring, the valve D is opened, allowing the whole of the soda solution to run via the pipe E into the generator. When the solution has run from the caustic soda tank into the generator the valve D is closed, then the necessary quantity of ferro-silicon is placed in the hopper on the top of the generator and the lid of the hopper closed, making a gas-tight joint. In small plants a little mineral grease is added to the generator, via the grease box.

The plant is then ready for operation, and silicol is cautiously fed into the generator by means of the hand-operated feed worked from F.

During the generation the fluid charge in the generator is kept stirred by means of the stirring mechanism worked from G. The hydrogen produced passes through the tube condenser (where it is cooled and thus freed from steam) and then on to the gas holder.

An excessive pressure, due to rapid generation of hydrogen, is guarded against by means of a water seal as shown.

When generation is complete, the resulting sodium silicate solution is rapidly run out via the trapped discharge pipe and the interior of the generator washed with cold water supplied from the tap B. Thermometers at T1, T2, T3, and T4 enable the temperature at different parts of the apparatus to be observed and, if necessary, controlled.

The description of the apparatus has, of necessity, to be somewhat general, as these plants are made in sizes varying from 1500 to 60,000 cubic feet per hour production and consequently differ in detail; thus, in large plants, the tube condenser is not employed and the hot hydrogen passes up a tower packed with coke, down which water is falling. Further, in large plants, the generator itself is water-jacketed, as the heat of chemical reaction would otherwise be excessive.

The silicol process has the advantage of giving a very great hydrogen production per hour from a plant of small cost—its disadvantage is that at the prevailing cost of the reagents employed the hydrogen is expensive.

To sum up, this process is exceedingly useful where large quantities of hydrogen are from time to time required, but it is not the best process to use where there is a constant hour-to-hour demand for hydrogen.

The Silicon Content of the Ferro-Silicon.—The grade of ferro-silicon used in this process is very important, as low-grade material does not yield anything like the theoretical quantity of hydrogen which should be obtained from the silicol present. This arises to a slight extent from the protective action of the impurities, which enclose particles of silicon and therefore prevent the caustic soda from attacking it.

The curve (Fig. 4), obtained experimentally, shows that to get even moderate efficiency ferro-silicon of over 80 per cent. silicon content should be used.

The Degree of Fineness of the Ferro-Silicon.—The degree of subdivision of the ferro-silicon is also important, not so much because of its effect on the total yield of hydrogen, but because of its influence on the rapidity of generation.

Fig. 5 indicates the speed of evolution of hydrogen from two samples of the same material, under identical conditions, except that one sample was much coarser than the other.

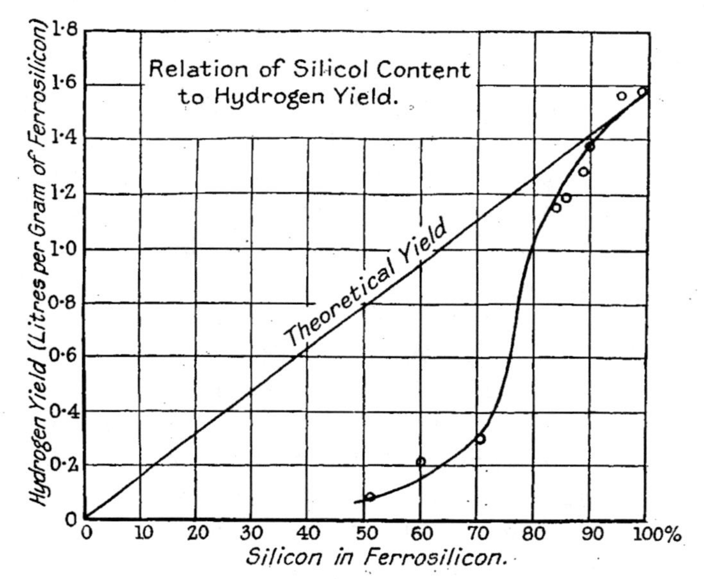
Relation of Silicol Content to Hydrogen Yield.
Theoretical Yield
Hydrogen Yield (Litres per Gram of Ferrosilicon)
1·8
1·6
1·4
1·2
1·0
0·8
0·6
0·4
0·2
0
0 10 20 30 40 50 60 70 80 90 100%
Silicon in Ferrosilicon.

FIG. 4.

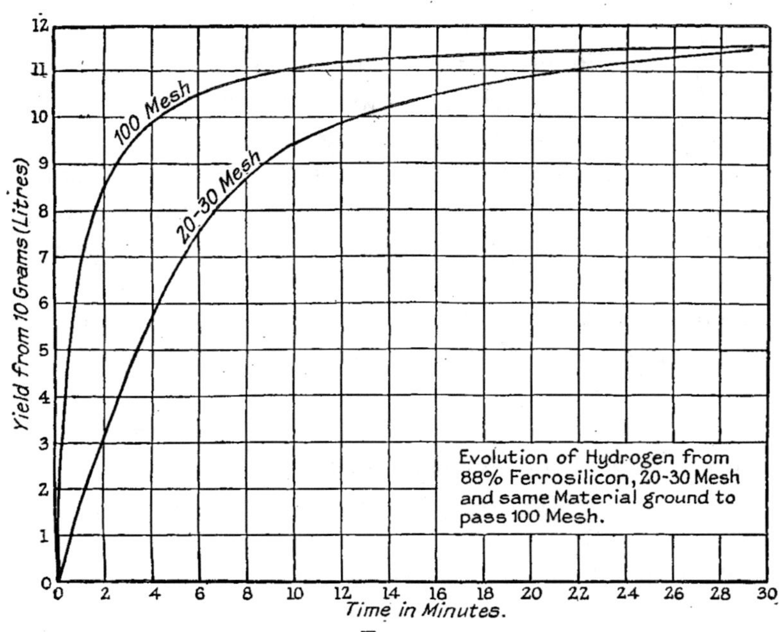
100 Mesh
20-30 Mesh
Evolution of Hydrogen from 88% Ferrosilicon, 20-30 Mesh and same Material ground to pass 100 Mesh.
Yield from 10 Grams (Litres)
12
11
10
9
8
7
6
5
4
3
2
1
0
0 2 4 6 8 10 12 14 16 18 20 22 24 26 28 30
Time in Minutes.

FIG. 5.

The Strength of the Caustic Soda.—The strength of the caustic soda is very important in this process. If the solution is too dilute, a very poor yield of hydrogen is obtained, and also another difficulty is introduced. When the caustic soda solution is very weak, on the introduction of the ferro-silicon or "silicol" reaction takes place, but the whole solution froths violently, the froth being carried along the pipes from the generator, causing trouble to be experienced in the valves, and tending to ultimately block the pipes themselves. On the other hand, the caustic soda solution may be too strong. In this case, before the whole of the caustic soda has reacted with the requisite amount of silicol, the solution becomes either very viscous or actually solid, so a poor yield is obtained and the sludge cannot be got out of the generator without allowing it to cool down and then digging it out by manual labour.

The following laboratory experiments with ferro-silicon containing 92 per cent. silicon and caustic soda containing 98 per cent. of sodium hydroxide illustrate the effect of soda solutions of varying strength, and also the effect of varying ratios of pure silicon to pure sodium hydroxide. From these it will be seen that the most economical results are obtained when a 40 per cent. solution of caustic soda is employed and the ratio of silicon to sodium hydroxide is approximately 1 to 1·6.

In practice such strong solutions are not used, as, owing to the evaporation of a good deal of the water during the process, towards the end a degree of concentration would be reached which would prevent the sludge from being run out of the plant. A solution containing about 25 per cent. of caustic soda is found to give in practice very satisfactory results. Such a solution of commercial caustic soda, containing about 25 per cent. of pure sodium hydrate, has at 100° F. a specific gravity of approximately 1·32—a figure which is very useful to remember, as by means of a hydrometer a rapid check can be made of the caustic soda solution being prepared for use in the process.

The ratio of silicol to caustic soda should be such that the ratio of pure silicon to pure sodium hydrate is as 1 to 1·72, but this figure is capable of modification to a slight extent, depending on the temperature of the mixture, which is naturally higher in large plants than in small ones.

Experiment.	Strength of Soda Solution.	Ratio of Silicon to pure Caustic Soda.	Yield in Cubic Feet per lb. of Silicol. [1] (At 30″ Bar. and 60° F.).
1	10 per cent.	1 to 0·745	13·62
2	10 ,,	1 ,, 1·065	14·30
3	10 ,,	1 ,, 1·480	15·36
4	10 ,,	1 ,, 3·200	16·80
5	30 ,,	1 ,, 0·852	19·35
6	30 ,,	1 ,, 2·13	23·90
7	30 ,,	1 ,, 3·19	23·58
8	40 ,,	1 ,, 1·58	24·10
9	40 ,,	1 ,, 3·19	24·50

The Use of Slaked Lime instead of Caustic Soda.—The experiments already described indicate that to obtain hydrogen from ferro-silicon a base must be used to react with it. It therefore occurred to the author that the cost of the operation of the process might be reduced by the substitution of slaked lime for caustic soda.

[1] Theoretically the maximum possible yield under these conditions of temperature and pressure would be 25·4 cubic feet per lb. of silicol of this purity.

Laboratory experiments, using ferro-silicon containing 92 per cent. silicon and pure slaked lime, were made to see if the following reaction took place :—

$$Si + Ca(OH)_2 + H_2O = CaSiO_3 + 2H_2.$$

The results of these experiments indicated that when 1 part of ferro-silicon, 2·5 parts of slaked lime, and 10 parts of water were used, a yield of 1·53 cubic feet of hydrogen was obtained per lb. of ferro-silicon used. Theoretically, under the conditions of the experiment, 25·4 cubic feet should have been produced per lb. of ferro-silicon; consequently it can be safely concluded that without an external supply of heat the suspected reaction only takes place to a very limited extent.

Remembering that slaked lime will decompose sodium silicate, producing caustic soda and calcium silicate, in accordance with the following equation :—

$$Na_2SiO_3 + Ca(OH)_2 = 2NaOH + CaSiO_3,$$

it was thought that the following reactions might take place if both caustic soda and slaked lime were employed at the same time :—

$$(1)\ 2Si + 2NaOH + 3H_2O = Na_2Si_2O_5 + 4H_2.$$
$$(2)\ Na_2Si_2O_5 + Ca(OH)_2 = 2NaOH + CaSi_2O_5.$$

Since the caustic soda in the solution would be regenerated after it had reacted with the silicol, it would be available for reacting with yet more silicol, and would consequently reduce the quantity of caustic soda used in the process.

The following experiments, using a mixture of slaked lime and caustic soda, appear to indicate that the surmise was partly or wholly correct, for with approximately only half as much caustic soda as ferro-silicon a yield of almost 16 cubic feet of hydrogen per lb. of ferro-silicon was obtained. By contrasting these experiments with those already given, using caustic soda alone, it will be seen that the yield obtained from the caustic soda is much greater than that which would have been obtained were no lime present. Whether in operating on a large scale equally good results would be obtained has not yet been determined.[1]

Experiment.	Ratio of Silicon to pure Caustic Soda.	Ratio of Silicon to Lime.	Yield in Cubic Feet per lb. of Silicol at 30″ Bar. and 60° F.
1	1 to 0·426	1 to 1·52	14·95
2	1 ,, 0·426	1 ,, 2·72	15·95
3	1 ,, 0·426	1 ,, 3·04	15·23

The Chemical Composition of the Sludge.—The equations which have already been given indicate that the products of this process are hydrogen and sodium disilicate in solution in water. Since, however, neither the ferro-silicon nor caustic soda employed are pure, in the practical production of hydrogen by this method, products other than those shown in the equations are found.

The commercial caustic soda employed always contains a certain amount of carbonate of soda, which takes no part in the reaction and is found unaltered in the sludge. The same remark applies to the iron contained in the ferro-silicon.

The following analysis gives the chemical composition of the sludge produced when 1414 lb. of ferro-silicon, containing 84 per cent. of silicon, and 2688 lb.

[1] Since the above experiments the author has found that there is a patent for the use of lime in conjunction with caustic soda and silicon, which, under the name of "Hydrogenite," has been employed by the French Army for inflating observation balloons in the field.

of caustic soda, containing 76 per cent. of sodium hydrate, was employed. Besides the sludge 29,300 cubic feet of hydrogen was produced, measured at a temperature of 60° F. and a barometer of 30 inches.

CHEMICAL COMPOSITION OF SLUDGE.

	Per Cent. by Weight.
Moisture	27·74
Silica	36·79
Sodium carbonate	6·04
Soda (Na_2O, other than carbonate)	20·08
Insoluble and undetermined	9·35
	100·00

CHEMICAL COMPOSITION OF SODA SOLUTION USED.

Caustic soda	24·5
Sodium carbonate	3·0
Specific gravity at 100° F.	1·324

CHEMICAL COMPOSITION OF FERRO-SILICON USED.

Silicon	84·0
Iron	6·9
Aluminium	5·3
Carbon	·2
Undetermined	3·6
	100·0

SCREEN [1] ANALYSIS OF FERRO-SILICON.

				Grms.	Per Cent.
Through	20,	on	30	141·5 =	10·28
,,	30	,,	40	85·0 =	6·17
,,	40	,,	50	98·0 =	7·11
,,	50	,,	60	85·5 =	6·21
,,	60	,,	70	103·0 =	7·48
,,	70	,,	80	117·0 =	8·49
,,	80	,,	90	64·0 =	4·64
,,	90	,,	100	104·0 =	7·55
,,	100	,,	120	81·0 =	5·88
,,	120	,,	150	338·0 =	24·55
,,	150	,,	200	89·0 =	6·46
Passing			200	70·0 =	5·12
				1376·5	99·94

[1] Standard I.M.M. screens.

The Use of Mineral Grease.—To reduce the frothing and priming in this process it is customary to introduce a little mineral grease, which floats on the surface of the caustic soda solution and prevents the formation of the froth to a considerable extent.

About 32 lb. of mineral grease are advocated per 1000 lb. of silicol used. However, if the caustic soda solution is strong, i.e. about 25 per cent. sodium hydrate, and the generator is wide, giving a large surface and a shallow depth to the caustic soda solution, no grease need be used at all.

Precautions to be Observed.—In this process very great care must be taken in the introduction of the ferro-silicon. When the ferro-silicon is attacked by the caustic soda large quantities of heat are given out, raising the temperature of the caustic soda solution. If the caustic soda solution is cold, ferro-silicon can be introduced into the solution far more rapidly than it is attacked by the soda; consequently there is likely to exist an accumulation of ferro-silicon in the solution, the temperature of which is gradually rising. A certain critical temperature is ultimately reached when the whole of the accumulated ferro-silicon is almost instantly attacked, with large yield of hydrogen and consequently the production of high pressure in the generator. Several explosions have been caused in this country from this reason. While it is impossible to give any definite figures, in the ordinary commercial plant for the production of hydrogen by this process the ferro-silicon should be added in small quantities, with a period of waiting between each addition, until the caustic soda solution reaches a temperature of about 180° F. At this temperature, with a 25 per cent. solution of caustic soda, high-grade ferro-silicon is almost instantly attacked, so it can then be added continuously at a rate which does not produce a pressure above the working pressure of the plant.

In plants for the operation of this process no red or white lead whatever should be used for making joints, as both these substances at a comparatively low temperature are reduced to metallic lead by ferro-silicon, with the evolution of large quantities of heat and the production of incandescence, the reaction taking place with such rapidity as to constitute almost an explosion. This can be easily illustrated by making a mixture of finely divided ferro-silicon and dry red lead, in which the ratio of the two is 1 part of pure silicon to 12·2 parts of red lead. If a match or the end of a cigarette is put to this mixture it goes off violently, with the production of great heat, in accordance with the following equation:—

$$Pb_3O_4 + 2Si = 3Pb + 2SiO_2.$$

That the temperature produced is exceedingly high can be well illustrated by putting, say, half an ounce of an intimate dry mixture of ferro-silicon and red lead, in which the proportions of the active principals are as indicated in the above equation, on a sheet of thin aluminium, say $\frac{1}{16}$ of an inch thick. On putting a match to this mixture it will be found a hole is melted in the aluminium sheet.[1]

All air must, if possible, be excluded from the plant prior to the introduction of the caustic soda, as otherwise in the early stages of generation of hydrogen an explosive mixture might arise which on ignition would

[1] The melting point of aluminium is 658° C.

produce a dangerous explosion. Such ignition might arise from a spark produced from the mechanism inside the generator, by ferro-silicon coming in contact with red lead which might have been used in making the joints in the plant, by incandescence produced by the reaction of ferro-silicon with caustic soda itself. If an intimate mixture of powdered caustic soda is made with ferro-silicon in the ratio indicated in the equations on the silicol process, and this mixture is just moistened with water, hydrogen is rapidly evolved and the reacting mass frequently becomes incandescent. Such conditions might arise in a plant for operation of this process, by the caustic soda being splashed on to some recess in the generator, there becoming concentrated, and ferro-silicon coming into contact with this concentrated solution. It is for this reason that caustic soda and ferro-silicon should never be stored in close proximity to each other, as this dangerous reaction may arise from the breaking of drums containing the two reagents.

Since this process is generally operated in conjunction with a gas-holder, the most easy way to exclude air is to allow hydrogen from the gas-holder to blow back through the plant prior to putting this in operation. Hydrogen equal to about four times the volume of the plant is required to thoroughly exclude the air.

The following patents with regard to this process are in existence :—

Consort. Elektrochem. Ind.—English patent 21032, September 14th, 1909.

French patent 418946, July 18th, 1910.

English patent 11640, May 13th, 1911.

Jaubert—French patent 430302, August 6th, 1910.

THE HYDROGENITE PROCESS.

There is a modification of this method known as the hydrogenite process whereby the use of an aqueous solution of caustic soda is avoided.

An intimate mixture of ferro-silicon and powdered caustic soda or lime is packed in strong cylinders communicating with a high pressure storage. By means of a fuse the temperature is locally raised so that chemical reaction takes place, with the production of hydrogen and sodium and calcium silicates.

This modification is covered by Jaubert, English patent 422296, 1910; English patent 153, 1911.

With Carbon.—If caustic soda is heated to dull redness with charcoal or anthracite or some other form of pure carbon, hydrogen is evolved and sodium carbonate and sodium oxide produced, in accordance with the following equation :—

$$4NaOH + C = Na_2CO_3 + Na_2O + 2H_2.$$

Theoretically, to produce 1000 cubic feet of hydrogen at 30 inches barometric pressure and 40° F. by this process, 222 lb. of caustic soda and 16·61 lb. of carbon are required, or a total weight of pure reagents equal to 238·61 per 1000 cubic feet of hydrogen produced.

The hydrogen produced by this process would be liable to contain traces of methane, arsine, and sulphuretted hydrogen, the amount depending on the purity of the coal used.

A modification of this process, whereby the caustic soda is replaced by slaked lime, is covered by a patent taken out in U.S.A. by Bailey in 1887.

With a Formate or Oxalate.—If sodium formate is

heated with caustic soda in the form of soda lime, the following reaction takes place :—

$$H.COONa + NaOH = Na_2CO_3 + H_2.$$

This method has been used for the production of hydrogen in the laboratory ; however, it cannot be regarded as singularly convenient.

If the sodium formate is replaced by sodium or potassium oxalate a similar reaction takes place :—

$$Na_2C_2O_4 + 2NaOH = 2Na_2CO_3 + H_2.$$

This last method, it is interesting to note, was employed by Amagat for the preparation of the hydrogen for his classic experiments on the relationship of pressure to volume.

(3) Methods in which Hydrogen is Derived from Water.

Hydrogen can be derived from water by means of the alkali and alkali earths groups of metals, but since all these are expensive, the production of hydrogen from these sources is limited to the requirements of the chemical laboratory.

With Lithium.—If metallic lithium is placed in water it is attacked by it, in accordance with the following equation, with the production of hydrogen and lithium hydrate :—

$$2Li + 2H_2O = 2LiOH + H_2.$$

It is interesting to note that since metallic lithium has a density of ·59 (the smallest density of any solid), it floats on the surface of the water while it is being attacked.

With Sodium.—If metallic sodium is placed in water it is attacked by it, in accordance with the following equation, with the production of hydrogen and sodium hydrate :—

$$2Na + 2H_2O = 2NaOH + H_2.$$

Since considerable quantities of heat are given out when the sodium is attacked by the water, much of which heat is communicated to the metal, it frequently melts while being attacked, the melting point of the metal being 95·6° C.

While the above equation expresses the principal reaction which takes place, a second reaction also occurs, leading to the production of sodium hydride, which is somewhat unstable under these conditions and occasionally explodes with violence, to avoid which a piece of apparatus has been designed by J. Rosenfeld.[1]

With Potassium.—If metallic potassium is placed in water it is attacked, in accordance with the following equation, with the production of hydrogen and potassium hydrate :—

$$2K + 2H_2O = 2KOH + H_2.$$

Such is the heat which is liberated during the reaction that if a piece of potassium is placed in a bucket of water, the metal is carried to the surface by the vigorous stream of hydrogen produced, and there becomes so hot as to ignite the hydrogen evolved.

The same remarks which have been made as to a secondary reaction with regard to sodium, apply with like force to potassium.

[1] "Prakt. Chem.," 48, 599-601.

With the Alkaline Earths.

With Magnesium.—If metallic magnesium is placed in water which is heated to its boiling point, hydrogen is slowly evolved, in accordance with the following equation, producing hydrogen and magnesium hydrate :—

$$Mg + 2H_2O = Mg(OH)_2 + H_2.$$

To accelerate the reaction, the magnesium is gently heated in a tube and steam passed over it.

With Calcium.—If metallic calcium is placed in water at ordinary atmospheric temperature it decomposes it in accordance with the following equation, producing a vigorous stream of hydrogen and calcium hydrate :—

$$Ca + 2H_2O = Ca(OH)_2 + H_2.$$

With Strontium.—A similar reaction to that already given for calcium takes place, but somewhat more vigorously.

With Barium.—A similar reaction to that already given for calcium takes place, but much more vigorously.

The Bergius Process.

With Iron.—If metallic iron is heated in the presence of water under very high pressure, hydrogen is evolved and magnetic oxide formed, in accordance with the following equation :—

$$3Fe + 4H_2O = Fe_3O_4 + 4H_2.$$

This method, which is known as the Bergius process, was put into commercial operation in 1913 at Hanover. It has two great advantages—a very pure hydrogen is produced, and since it is under great pressure, it can be used to fill bottles without the use of a compressor. The chemical composition of the hydrogen produced is given as follows :—

Hydrogen	99·95	per cent.
Carbon monoxide . . .	·001	" "
Saturated hydrocarbons . .	·042	" "
Unsaturated hydrocarbons .	·008	" "

but amount of the carbon compounds must be greatly influenced by purity of the iron employed; however, it appears to be a fact that little or no sulphuretted hydrogen is produced even if the iron contains appreciable quantities of sulphur.

While it has been stated that the hydrogen is produced by the action of water at high temperature and pressure upon metallic iron, this does not entirely describe the chemistry of the process, for the inventor has found that the presence of certain metallic salts in the solution, and also other metals, greatly increase the speed of production of hydrogen. The following comparative table of productions of hydrogen from equal weights of iron clearly illustrates this point :—

	Temperature °C.	Volume of Hydrogen per Hour.
Iron and pure water . . .	300	230 c.c.
Iron, water, and ferrous chloride	300	1390 "
Iron, water, ferrous chloride, and metallic copper . . .	300	1930 "
Iron, water, ferrous chloride, and metallic copper . . .	340	3450 "

In the commercial employment of this process it is believed that the working pressure is about 3000 lb. per sq. inch and the temperature 350° C.

In the discharge from the vessel in which the hydrogen is produced there is a reflux condenser which effectively prevents any steam from escaping from the plant when the hydrogen is drawn off.

One of the remarkable features of this process is the fact that since the water pressure is so high, it penetrates right into the iron particles; consequently they are entirely employed, and a yield closely approaching the theoretical is obtained.

Theoretically, 1000 cubic feet of hydrogen (at 30 inches barometer and 40° F.) would be produced with an expenditure of 116·5 lb. of metallic iron.

The size of the plant is very small for the yields obtained, it being stated that a generator of 10 gallons' capacity gives 1000 cubic feet of hydrogen at atmospheric temperature and pressure per hour.

After the completion of the reaction the pressure can be let off from the generator and the magnetic oxide of iron produced removed and reduced by water gas to the metallic state, when it can again be employed in the process.

It is claimed that the cost of hydrogen by this process is exceedingly low; consequently, if this is correct and the mechanical difficulties of manufacturing generators to withstand the very high pressures and chemical action are satisfactorily overcome, this process would appear to be one of the highest value for the commercial production of hydrogen.

A certain amount of information with regard to this process can be found in the following paper: "Production of Hydrogen from Water and Coal from Cellulose at High Temperatures and Pressures," by F. Bergius, the "Journal of the Society of Chemical Industry," vol. xxxii., 1913.

The process is protected by the following patents :—

German Patent, 254593, 1911.

French Patent, 447080, 1912.

English Patents, 19002 and 19003, 1912.

United States Patents, 1059817, 1059818, 1913,

all in the name of F. Bergius.

While the previous method is of commercial importance, the following method is of interest :—

When the ordinary process of rusting of iron takes place, hydrogen is evolved.

It is generally considered that iron does not rust when it is in contact with perfectly pure water, free from carbon dioxide or any other mild acid. In the following method iron filings are placed in a steel bottle partly filled with water saturated with carbon dioxide; the bottle is then closed and sealed. It is then agitated, the following reaction taking place :—

$$Fe + H_2O + CO_2 = FeCO_3 + H_2.$$

This method is one of interest and is described by Bruno in the "Bull. Soc. Chim.," 1907, [iv.], 1661. It cannot, however, be regarded as having a commercial use.

With Hydrides.—As has already been stated, the hydrides of the metals of the alkali and alkaline earth groups produce hydrogen on being placed in water. However, in only two cases are these reactions worth consideration.

With Lithium Hydride.—If lithium hydride is brought into contact with water, hydrogen is evolved and lithium hydrate formed, in accordance with the following equation :—

$$Li_4H_2 + 4H_2O = 4LiOH + 3H_2.$$

Such is the rarity of lithium at the present time that the above is neither a commercial nor a laboratory method of producing hydrogen. It is, however, of the greatest interest, owing to the large yield of hydrogen obtained from a small weight of lithium hydride. Theoretically, 1000 cubic feet of hydrogen at 30 inches barometer and 40° F. are produced from 27·76 lb. of pure lithium hydride and 66·6 lb. of water. Many ingenious ideas have been put forward for the employment of lithium hydride in airships so that in the event of an airship loosing gas from some cause, this may be replaced by hydrogen manufactured in the airship. As has been seen, theoretically, 94·36 lb. of reagents are required to produce 1000 cubic feet of hydrogen at 30 inches barometer and 40° F. Now this amount of hydrogen would have a lift of 74·06 lb., so if the products of the manufacture of hydrogen were dropped the buoyancy of the ship would be increased by 94·36 + 74·06 lb., or 168·42 lb. for every 94·36 lb. of material dropped from the ship. However, interesting as these suggestions are, such is the rarity and cost of lithium that at the present time they are not capable of realisation, though future discoveries of lithium minerals and cheaper methods for the production of lithium hydride may possibly render these ideas of practical value.

The Hydrolith Process.

With Calcium Hydride.—If calcium hydride is brought into contact with water, hydrogen is evolved and calcium hydrate formed, in accordance with the following equation :—

$$CaH_2 + 2H_2O = Ca(OH)_2 + 2H_2.$$

Theoretically, to produce 1000 cubic feet of hydrogen at 30 inches barometric pressure and 40° F., 58·4 lb. of pure calcium hydride and 49·95 lb. of water are required, or a total weight of 108·4 lb. of pure reagents per 1000 cubic feet of hydrogen. Since, however, water does not have to be carried in most parts of Europe, the theoretical weight to be carried per 1000 cubic feet of hydrogen required is 58·4 lb. This method, known as the Hydrolith process, has been satisfactorily employed by the French Army in the field for the inflation of observation balloons, the calcium hydride being packed in air- and water-tight boxes for transportation. In the commercial production of calcium hydride small quantities of calcium nitride are produced, which, when the hydride is attacked with water gives rise to ammonia, in accordance with the following equation :—

$$Ca_3N_2 + 6H_2O = 3Ca(OH)_2 + 2NH_3.$$

However, as ammonia is very readily soluble in water, if the hydrogen produced in the process is scrubbed with water the ammonia is almost entirely removed and an exceedingly pure hydrogen results.

This process is protected by a French patent, No. 327878, 1902, in the name of Jaubert.

With Metallic Sodium and Aluminium Silicide.— If a mixture of metallic sodium and aluminium silicide is placed in water, hydrogen is evolved, with the production of sodium silicate and aluminium hydrate, in accordance with the following equation :—

$$Al_2Si_4 + 8Na + 18H_2O = Al_2(OH)_6 + 4Na_2SiO_3 + 15H_2.$$

Theoretically, 1000 cubic feet of hydrogen at 30 inches barometer and 40° F. are produced from 64·8 lb. of this mixture. It is, however, believed that the practical yield is about 80 per cent. of this figure.

This process is essentially one for the production of hydrogen for war purposes, though the author does not know of any actual use of it. The mixture can be made into briquettes, which must be packed into air- and water-tight boxes. The method, which is sometimes known as the "Sical process," is protected by a United States patent—977442, 1910—in the name of Foersterling and Philipps.

With Aluminium.—If ordinary metallic aluminium is placed in even boiling water, little or no chemical action takes place. However, if the aluminium is first amalgamated with mercury it is rapidly attacked by hot water, with the formation of aluminium hydrate and hydrogen, in accordance with the following equation :—

$$2Al + 6H_2O = Al_2(OH)_6 + 3H_2.$$

Theoretically, to produce 1000 cubic feet of hydrogen at 30 inches barometric pressure and 40° F., 50 lb. of aluminium are required.

In the commercial application of this method it is not necessary to amalgamate the metallic aluminium with mercury by hand, as advantage is taken of the fact that aluminium will reduce aqueous solutions of salts of mercury to the metallic state, in accordance with the following equation :—

$$2Al + 3HgCl_2 = 2AlCl_3 + 3Hg.$$

Consequently, if there is an excess of aluminium over that required by the equation, this excess will be automatically amalgamated by the metallic mercury as it is produced.

In a practical application of this method by Mauricheau Baupre,[1] fine aluminium filings are mixed with a small proportion of mercuric chloride ($HgCl_2$) and potassium cyanide (KCN), which causes a slight rise in temperature and produces a coarse powder, which is quite stable if kept free from moisture. This mixture can be kept in air- and water-tight boxes until it is required, when it can be gradually added to water kept at about 70° C. A brisk evolution of hydrogen then takes place which closely approximates to the theoretical yield.

Another very interesting application of this increased chemical activity of aluminium when amalgamated with mercury is incorporated in a toy which is sometimes seen on sale under the name of "Daddy Tin Whiskers". This toy consists of an aluminium stamping of a face and a pencil, the core of which is filled with a preparation chiefly composed of a mercury salt. It is operated by rubbing the eyebrows and chin with this special pencil. Shortly afterwards white hairs of aluminium oxide (Al_2O_3) gather wherever the pencil has touched the aluminium.

To operate the above process for the manufacture of hydrogen it is necessary that the aluminium should be as pure as possible and should not contain copper. The commercial light alloy known as "duralumin," which contains about 94 per cent. of aluminium and 4 per cent. of copper, is entirely unsuitable for generating hydrogen in the method above described, as it is almost unattacked by even boiling water containing a small quantity of a mercury salt.

[1] "Comptes Rend.," 1908, 147, 310-1.

The following patents deal with this process :—

French patent 392725, 1908, in the name of Mauricheau Baupre.

English patent 3188, 1909, in the name of Griesheim.

German patent 229162, 1909, in the name of Griesheim.

With Aluminium Alloy.—The following alloy—

Aluminium	78-98 parts.
Zinc	15-1·5 „
Tin	7-0·5 „

is made and cast into a plate; after cooling it is amalgamated with mercury. After amalgamation the plate is heated as strongly as possible without volatilising the mercury. When it has become thoroughly amalgamated it is allowed to cool and is then ready for use.

If this alloy is put into hot water it readily yields hydrogen; the hydrogen yield is proportionate to the aluminium and zinc content.

The gas produced is very pure.

This process is protected by the following patent: Uyeno, British patent, 11838, 1912.

With Steam.

In considering the production of hydrogen from steam, a considerable number of processes must be considered in which the first stage (which is common to all the processes) consists in the manufacture of blue water gas; consequently, prior to the description of these processes, amongst the most important of which are :—

The Iron Contact process,
The Badische process,
The Linde-Frank-Caro process,
the manufacture of water gas will be described.

The Manufacture of Water Gas.

When steam is passed over red-hot carbon, the two following chemical reactions take place :—

$$(1)\ C + H_2O = CO + H_2.$$
$$(2)\ C + 2H_2O = CO_2 + 2H_2.$$

The question as to which equation represents the predominant reaction taking place, depends on the temperature of the carbon ; roughly speaking, the higher the temperature the more closely does the reaction coincide with the first chemical equation.

The following experimental results (H. Bunte, "J. für Gasbeleuchtung," vol. 37, 81) clearly illustrate the effect of temperature on the chemical composition of the products of the reaction :—

Temperature °C.	Per Cent. of Steam Decomposed.	Composition, by Volume of Gas Produced.		
		H_2.	CO.	CO_2.
674	8·8	65·2	4·9	29·8
758	25·3	65·2	7·8	27·0
838	41·0	61·9	15·1	22·9
954	70·2	53·3	39·3	6·8
1010	94·0	48·8	49·7	1·5
1125	99·4	50·9	48·5	0·6

In the first of the chemical equations given, it will be seen that the products are composed of 50 per cent.

hydrogen and 50 per cent. carbon monoxide, while in the second, the composition is 66·66 per cent. hydrogen and 33·33 per cent. carbon dioxide; in Dr. Bunte's experiments, figures closely approximating to the first equation were obtained when the temperature of the carbon was 1000°-1100° C., while figures similar to the products indicated in the second equation were found when the temperature was 674° C.

Now, whatever purpose water gas may be required for, its use for this purpose depends on the fact that the gas will combine with oxygen with the evolution of heat, consequently the plant should be worked to make the product with the highest calorific power for the lowest fuel consumption. This requirement is reached more closely if the plant is operated so that the first equation represents the chemical reaction which takes place; consequently, in the practical manufacture of water gas the coke or other fuel in the gas producer should be at a temperature of about 1000° C.

The chemical reaction producing the decomposition of the steam is endothermic, that is to say, as the reaction proceeds, the temperature of the coke falls, so that in order to obtain a gas approximating to the products in the first equation, heat must be supplied to the coke, to counteract the fall in temperature, due to its reaction with the steam.

In the oldest type of plant, the coke which was used for the manufacture of the water gas was in a cylinder, which was externally heated by a coke or coal fire; however, this procedure was not very efficient, and the practice is not in use at all at the present time.

In practice to-day there are two methods of making water gas, one the English or Humphrey and Glasgow method, and the other the Swedish or Dellwick-Fleischer method.

English Method.—It has already been pointed out that from thermal chemical reasons, the coke through which the steam is passing in the manufacture of water gas should be at about 1000° C. in order to obtain good results, and that as a result of the reaction between the coke and steam, the temperature of the former falls, necessitating the addition of heat to the coke mass, in order to keep up the efficiency of the process.

It is in the method of maintaining the coke temperature that the English and Swedish systems differ. In both systems the coke is kept at the proper temperature by shutting off the steam supply from time to time, and blowing air through the coke, the products of the air blast passing out of the generator through a different passage to those of the steam blast.

The effect of blowing air through the coke is of course to produce heat, for the following reactions to a lesser or greater extent take place:—

$$(1)\ C + O_2 = CO_2,$$
$$(2)\ CO_2 + C = 2CO,$$

and the heat, which is produced by the combustion of some of the coke, heats the remainder, thus raising its temperature, so that the air blast can be shut off, and the steam blast again turned on.

In the English system the depth of the coke in the generator is considerable, consequently the carbon dioxide formed at the base of the fire tends to be reduced in the upper part of the fire by the hot coke, in accordance with equation (2), therefore in the English system during the air blast a combustible gas is produced.

However, while at first sight this might appear to be an advantage, there are several disadvantages associated with this method of working. In the first place, the gas which is produced during the air blast, though combustible, is not a gas of high calorific power, as it contains such a large amount of atmospheric nitrogen ; in fact, under the most favourable circumstances, the gas produced during the air blast will not contain over 30 per cent. of carbon monoxide, while the rest of it will be incombustible, being chiefly nitrogen together with some carbon dioxide. Another disadvantage of this system is that since the coke is permanently burnt only to carbon monoxide, the amount of heat actually generated in the coke mass is comparatively small, consequently the rate of temperature rise in the coke mass is slow.

In the *Swedish* or *Dellwick-Fleischer* method, the coke temperature is from time to time raised by means of an air blast, but in this case the depth of fuel is relatively shallow, so that the carbon burnt remains permanently in the form of carbon dioxide ; and since in burning equal weights of carbon to carbon monoxide and carbon dioxide over three times as much heat is generated *in situ* when the carbon is burnt to carbon dioxide than when burnt to carbon monoxide, the rate of rise of temperature of the coke mass in the generator is much more rapid than is the case in the English system, and consequently the period occupied by the air blast is very much reduced.

Fig. 6 shows a diagram from Dellwick's English patent 29863, 1896, illustrating his plant for the production of blue water gas.

A is the generator provided with a coke receptacle B, which passes through a stuffing-box D placed on the cover or top of the generator. The object of this receptacle is to keep the fuel height constant in the generator; if B is kept filled with fuel, as that which is on the grate burns away, fresh fuel will run in from B and will keep the depth of fuel on the grate constant.

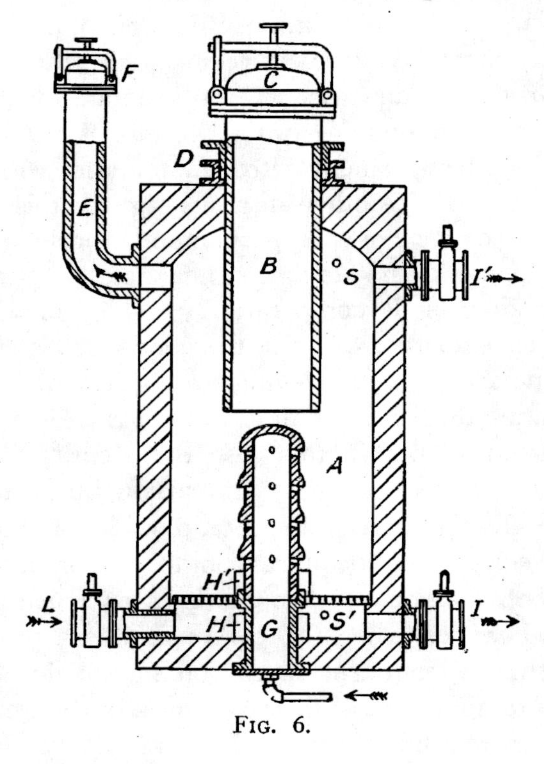

FIG. 6.

L is the main air inlet, while a secondary supply of air takes place through the vertical pipe G, the purpose of this latter air inlet being to ensure a thorough supply of air to all parts of the fuel bed.

S and S′ are steam inlets, I and I′ are gas outlets, and E is an outlet for the products made during the air blast.

The method of working this generator would be as follows :—

When a coke or other fire of suitable depth has been obtained on the grate, the receptacle B is charged with fuel, and the lid C firmly closed ; valves I and I′ are closed and valve F opened, then air under suitable pressure is admitted through L and G ; this causes the fuel to burn with rise in temperature of the unburnt portion, while the products of combustion, containing about 20 per cent. of carbon dioxide and 70 per cent. of nitrogen, escape by the passage E.

When the temperature of the coke on the hearth has been raised to about 1000° C. the air blast is stopped, valve F closed, valve I′ opened, and steam admitted through S′ with the consequent production of blue gas, which passes out to a scrubber and holder, via the valve I′.

When as a result of the decomposition of the steam by the fuel mass, the temperature of the latter has fallen below the economic limit, the steam supply is shut off, and the air blast started again to raise the fuel temperature. When the temperature is again suitable, the air is shut off and steam again passed through the fuel, but on this occasion downwards from the steam supply S, the water gas passing out by the outlet I.

The object of this alternation of the direction of the steam blast is to keep the temperature as uniform as possible throughout the fuel mass.

Fig. 7 shows a modern water gas producer, which is self-explanatory ; the fuel charging is done after every third steam blast, and the depth of the fuel kept correct by means of a gauge rod, dropped through the lid at the top of the generator. The same alternation in the direction of the steam blast is maintained, while during the air blast the products of combustion escape through the lid at the top of the generator, which is open during this stage.

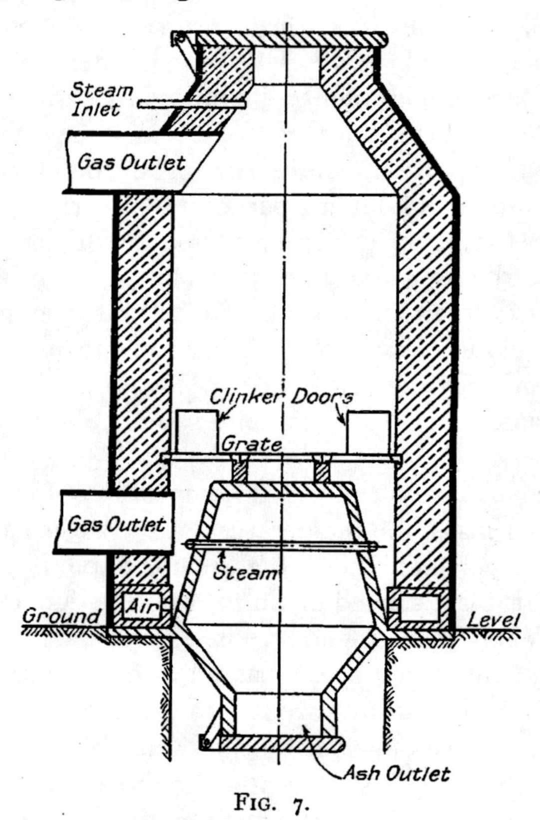

FIG. 7.

The sequence of operation with a standard generator, having a circular hearth about 5 feet 6 inches in diameter, would be :—

1. Air blast . . . . 2 minutes
2. Steam up . . . . 6 ,,
3. Air blast . . . . 1 minute

4. Steam down . . . . 6 minutes
5. Air blast . . . . 1 minute
6. Steam up . . . . 6 minutes.

At the end of the last operation, additional fuel would be added and the sequence again started. The air supply main would be at a pressure of about 15 inches of water above the atmospheric, while the steam main would be at about 120 lb. per sq. inch, the rate of flow of the steam being about 45 lb. per minute, during the steaming periods.

Working under the conditions described, using coke of the following composition as a fuel :—

	Per Cent.	
Moisture	6·0	by weight.
Ash	9·0	,,
Volatile sulphur	1·35	,,
Nitrogen	0·6	,,
Carbon, etc. (by difference)	83·05	,,
	100·00	

a water gas of about the following composition would be obtained :—

	Per Cent.	
Hydrogen	52·0	by volume.
Carbon monoxide	39·6	,,
Methane	0·4	,,
Carbon dioxide	3·5	,,
Sulphuretted hydrogen	0·5	,,
Nitrogen	4·0	,,
	100·0	

for a consumption in the generator of about 35-40 lb. of coke per 1000 cubic feet of blue water gas measured at atmospheric temperature and pressure.

A consideration of this coke consumption is instructive; from the analysis of the water gas, it will be seen in 1000 cubic feet of the gas there are—

396 cubic feet of carbon monoxide.
35 " " " " dioxide.

If the barometer is 30 inches and the temperature 60° F., 1000 cubic feet of carbon monoxide weighs 74·6 lb.

$\therefore$ 396 " " " " " $\frac{74{\cdot}6 \times 396}{1000}$
$= 29{\cdot}6$ lb.

But carbon monoxide contains $\frac{12}{28}$ of its total weight of carbon.

$\therefore$ 396 cubic feet of carbon monoxide contains $\frac{29{\cdot}6 \times 12}{28}$ lb. carbon = 12·7 lb.

Similarly,
1000 cubic feet of carbon dioxide weighs 117·3 lb.

$\therefore$ 35 " " " " " $\frac{35 \times 117{\cdot}3}{1000}$
$= 4{\cdot}1$ lb.

But carbon dioxide contains $\frac{12}{44}$ of its total weight of carbon.

$\therefore$ 35 cubic feet of carbon monoxide contains $\frac{4{\cdot}1 \times 12}{44}$ lb. of carbon = 1·1 lb.

Adding these two results together, it is seen that while 35-40 lb. of coke, equivalent to 29-33 lb. of carbon, are consumed in the generator per 1000 cubic feet of water gas, only 13·8 lb. of this carbon, or 42-46 per cent., are present in the gas produced, the bulk of the remainder of the carbon consumed in the generator being burnt during the air blast period, and the remainder lost in the ash pit, and during clinkering; however, while these figures are instructive, as indicating the magnitude of air blast consumption of fuel, to gain comparative figures it is necessary to obtain the calorific power of the coke consumed, and of that of the water gas produced from a given weight of coke.

If 35 lb. of the coke, the analysis of which has been already given, are consumed in the production of 1000 cubic feet of water gas at 30 inches barometer and 60° F., of the composition which has also been given, it will be found that the calorific power of the coke consumed, compared with that of the gas produced, is as

$$516 : 342,$$

that is to say, judged on a thermal efficiency basis, the efficiency of the producer working under these conditions is

$$\frac{342 \times 100}{516} = 66 \cdot 3 \text{ per cent.},$$

which is a figure such as is obtained in ordinary commercial water gas manufacture.

The analysis of the water gas so far given enumerates the chief constituents, but in reality there are traces of other products, such as carbon bisulphide, carbonyl sulphide, and thiophene, derived from the sulphur in the fuel, which, minute in quantity, may nevertheless in the certain chemical processes produce appreciable and undesirable results: from the iron contained in the fuel, minute amounts of iron carbonyl are formed, which in most processes in which water gas is used is a matter of no importance, but if the gas is to be used for lighting with incandescent mantles, its removal is desirable.

The producer, which has been described, is not in practice absolutely continuous in operation, as from time to time the process has to be interrupted in order to remove the clinker from the fire.

6

The process of "clinkering," besides requiring labour, is wasteful, as hot fuel as well as clinker is drawn from the fire, consequently various devices have been designed to make self-clinkering producers.

The majority of these designs consist essentially of a rotating conical hearth. Fig. 8 shows a device described in English patent 246111, 1909, which is almost self-explanatory. The clinker pan *h* and the blast nozzle *i* are connected and free to rotate on the ball race shown in the vertical section. The end of the blast nozzle *i* is fitted with helical excrescences with holes *k* for steam and air in their trailing edge. During the working of the producer, the nozzle and clinker pan are rotated, any clinker forming being broken up between the helical vanes on the fixed water jacketed body of the producer and those on the rotating blast nozzle. The clinker on being broken up falls into the clinker hearth, which is filled with water to such a depth as to make a water seal between the producer body and the moving hearth.

The bottom of the clinker hearth has fixed ribs, which tend to hold the crushed clinker, which during the rotation of the hearth is carried round until it is brought against the fixed vane *o*; this lifts it out of the water.

Producer hearths of the type described do not appear to effect any appreciable saving in fuel, but since they eliminate clinkering, they have a decided advantage, as the gas yield is greater in a given time than would otherwise be the case.

Purification of Water Gas.—For most industrial purposes, it is necessary that the crude water gas should

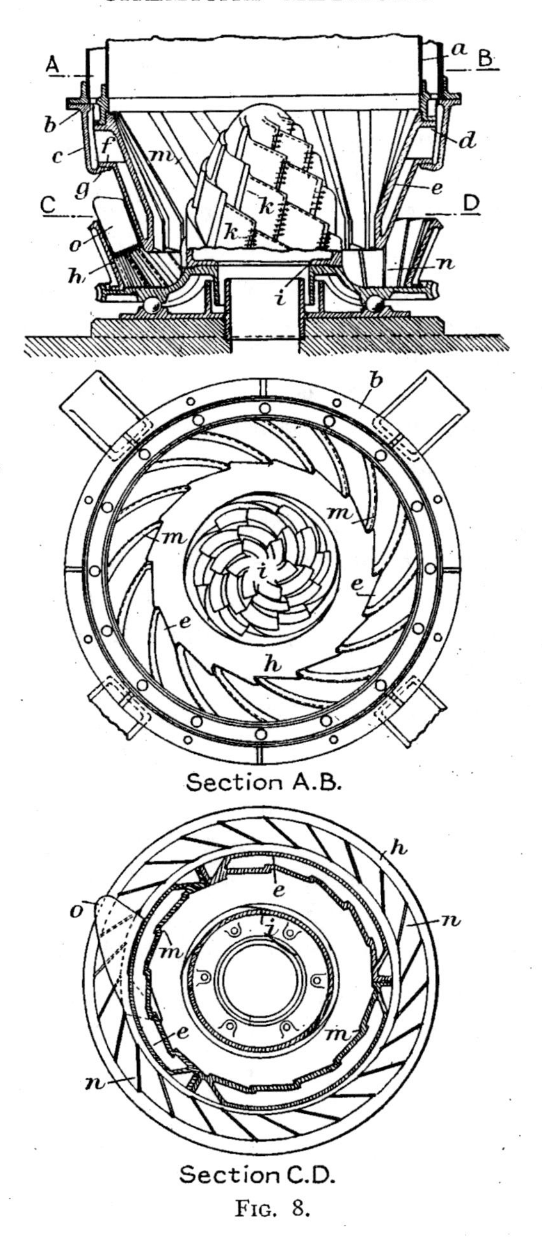

Section A.B.

Section C.D.

FIG. 8.

be purified before its ultimate use. For practically every process in which water gas is used it is necessary that it should be freed from the impurities which it mechanically contains, and which are composed of ash and dust, carried by the gas from the producer.

The mechanically retained impurities in water gas are removed by scrubbing the gas with water, that is to say, by passing it up a tower, down which water is falling. Not only does this water scrubbing remove the mechanically retained impurities, but it also, by reducing the temperature of the gas, causes the condensation and removal of the minute quantity of iron carbonyl contained in the gas.

Removal of Sulphuretted Hydrogen.—For most purposes for which water gas is required it is desirable that it should be free from sulphuretted hydrogen; this is usually accomplished by passing the gas at about 55°-65° F. over hydrated oxide of iron, when the following reaction takes place:—

$$Fe_2(OH)_6 + 3H_2S = 2FeS + 6H_2O + S.$$

After lapse of time, the hydrated ferric oxide ceases to have any sulphuretted hydrogen-absorbing power, so the gas is diverted through other hydrated oxide, and the spent oxide removed and placed in the open air, when, after moistening with water and exposure, the following reaction takes place:—

$$4FeS + 6H_2O + 3O_2 = 2Fe_2(OH)_6 + 4S.$$

Thus it is seen the original oxide can be reproduced, and on reproduction can be used for the absorption of fresh sulphuretted hydrogen. In practice each revivification increases the free sulphur content of the oxide about 7 per cent., and as time goes on the free sulphur in the iron oxide increases to 50-60 per cent. sulphur, when it commands a ready sale to manufacturers of sulphuric acid; roughly speaking, 1 ton of oxide will purify 2,000,000 cubic feet of gas before it is finally spent.

In this country, it is not generally necessary to heat the hydrated oxide of iron through which the crude water gas is passed, as the heat evolved by the chemical reaction is sufficient to keep the oxide at a suitable temperature. However, in many parts of the world, where the winter temperature is exceedingly low, it is necessary to pass steam coils through the oxide, as otherwise no absorption of sulphuretted hydrogen takes place.

The reason for this failure to absorb the sulphuretted hydrogen is due to the fact, already given in the equation, that with the absorption of the sulphuretted hydrogen, water is produced, which freezes on the surface of the hydrated iron oxide, and thus prevents further sulphuretted hydrogen coming in contact with it.

In the practical removal of sulphuretted hydrogen, it is desirable to have quite a considerable amount of water in the hydrated oxide (about 15 per cent. by weight), as this tends to keep it open and thus keep the pressure necessary to get the water gas through the oxide quite low; it is also desirable to keep the oxide alkaline, consequently about 1 per cent. of lime is mixed with it to accomplish this.

When new hydrated oxide is put in water gas purifiers, even though it may contain a sufficiency of water, it tends to cake together and create back pressure.

This can be prevented, either by mixing sawdust with the new oxide before putting it in the purifiers (about 1 part to 5 of oxide by volume) or by mixing some already used oxide containing a considerable amount of free sulphur with the new oxide; this also tends to prevent caking.

In ordinary commercial purification of water gas, 100 tons of hydrated ferric oxide will effectively purify 200,000 cubic feet of crude water gas per 24 hours; this allows of keeping 20-30 tons of "revivified" oxide in reserve, available to replace the working oxide as it becomes "spent".

This degree of purification of crude water gas to be used in the manufacture of hydrogen is common to all the processes using it; in some of the processes special methods of purification are employed, and these will be given in the description of the process which renders such methods necessary.

The Iron Contact Process.

Of all the processes for the production of hydrogen in which water gas represents one of the active reagents, the Iron Contact process is the most important, as it is by this process that the greater amount of the world production of hydrogen for use in industry and war is at present made; but important as this process is, it is doubtful if it will maintain its present pre-eminent position during the next few years, as other processes, more economical, but at present not so reliable, are already in existence, and with lapse of time greater reliability will probably be obtained in these later processes, which will result in the Iron Contact process occupying a less important position in hydrogen production than it does to-day.

When steam is passed over heated metallic iron,

hydrogen is produced in accordance with the following equation :—

$$3Fe + 4H_2O = Fe_3O_4 + 4H_2.$$

Theoretically, to produce 1000 cubic feet of hydrogen at 30 inches barometric pressure and 40° F., 116·5 lb. of iron and 49·95 lb. of steam are required : however, in practice these figures are not closely approached because the magnetic oxide of iron formed tends to shield the metallic iron from the action of the steam ; indeed, the reaction may be regarded as merely a surface one.

When the protective action of the magnetic oxide has reached such a degree that the yield of hydrogen has become negligible, the supply of steam is stopped, and the water gas is passed over the magnetic oxide, reducing it to metallic iron, in accordance with the following equations :—

$$Fe_3O_4 + 4H_2 = 3Fe + 4H_2O$$
$$Fe_3O_4 + 4CO = 3Fe + 4CO_2.$$

Then further steam can be passed over the iron, with the production of further hydrogen.

Thus, it is seen that the same iron is used continuously, and steam and blue water gas are the two reagents consumed. Such is the chemical outline of the Iron Contact process ; however, in practice, the process is somewhat more complex and very much less efficient than either the Electrolytic process or the Badische process, both of which are described at a later stage, nor can the hydrogen produced be regarded as so satisfactory for some industrial purposes, such as fat hardening, as that made by the other two processes.

In the practical working of the Iron Contact process, the process is not begun by passing steam over hot metallic iron, but by manufacturing the iron *in situ*, by reducing iron ore, such as hematite, with the water gas, which can be expressed by the following equations :—

$$2Fe_2O_3 + 3H_2 = 2Fe + 3H_2O$$
$$Fe_2O_3 + 3CO = 2Fe + 3CO_2.$$

The advantage of this procedure is that a spongy coating of metallic iron is obtained on the refractory iron oxide, with the result that the iron and the resulting magnetic oxide tend to be held together, and so keep the material open, and therefore free from back pressure to the passage of the steam and water gas.

In practice, to obtain a yield of 3500 cubic feet of hydrogen per hour, about 6 tons of iron ore are required. This ore, both in its original form and its subsequently surface altered state, is kept at a temperature of 650°-900° C. ; if lower than 650° C. the reactions become very slow, and if higher than 900° C. the material tends to frit, and become less open, thus creating resistance to the flow of gas and steam.

In the practical working of the Iron Contact process, the process consists of three stages :—

1. Reducing.
2. Purging.
3. Oxidising.

Reducing.—The reducing stage consists in passing water gas over the heated oxide, thus producing a coating of metallic iron on the oxide. During the first moment of reducing the reaction is comparatively effective, but with fewer opportunities for the gas to come into contact with unacted-upon oxide, the water gas is less and less effectively used, and consequently the gas on leaving the retorts contains more and more hydrogen and carbon monoxide as the reaction continues.

This variation in the efficiency of reduction, with lapse of time, is clearly illustrated in the graph, Fig. 9, which shows the carbon monoxide and carbon dioxide content of the water gas after passing at the rate of 9000 cubic feet per hour over 4·2 tons of iron oxide, heated to 750° C.

In practice, it is found that the speed of reduction is

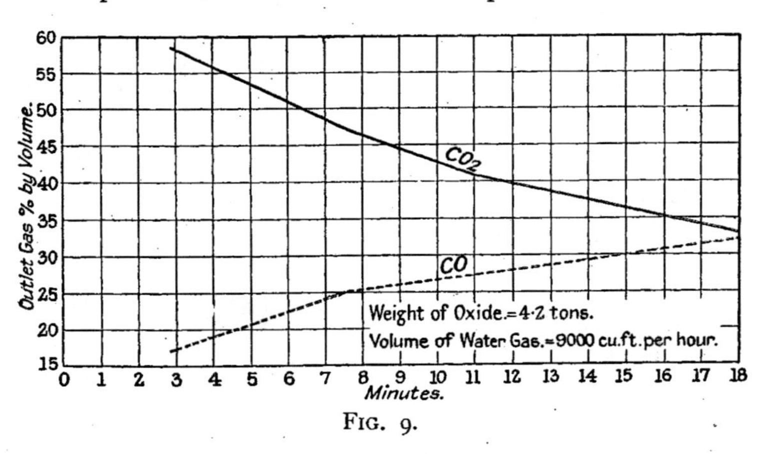

FIG. 9.

much slower than the speed of oxidation, consequently, in practice, the duration of the various stages is :—

Reducing	20 minutes.
Purging	35 seconds.
Oxidising	9 minutes, 25 seconds.

Purging.—When the reducing stage is stopped, the retort or retorts, containing the surface reduced oxide, is, or are, filled with an atmosphere of partly altered water gas; consequently when steam is turned on hydrogen is produced contaminated with the residual water gas; thus impure hydrogen is allowed to flow into some receptacle, where it is subsequently used for heating, or some other process, which will be described later.

At the end of 35 seconds the outflow of the gas is altered, and the now comparatively pure hydrogen is directed into a gas holder, or wherever it may be required.

Oxidising.—The oxidising stage is exactly the same as the purging stage, except as to the direction of outflow of the resulting hydrogen.

In normal working the gas produced has approximately the following composition :—

	Per Cent. by Volume.
Hydrogen	97·5
Carbon dioxide	1·5
„ monoxide	·5
Sulphuretted hydrogen	·03
Nitrogen (by difference)	·47
	100·00

Purification of Crude Hydrogen.—The crude hydrogen is first scrubbed with water, which besides removing mechanically contained impurities also reduces the amount of carbon dioxide, as this gas is soluble in water.

The hydrogen is then passed through boxes containing slaked lime, where both the carbon dioxide and sulphuretted hydrogen are absorbed in accordance with the following equations :—

$$Ca(OH)_2 + CO_2 = CaCO_3 + H_2O,$$
$$Ca(OH)_2 + H_2S = CaS + 2H_2O.$$

However, since there is no simple process of revivifying the lime after use, it is probably better practice to pass the crude hydrogen first through an iron oxide box, identical with that used in purifying water gas; here the sulphuretted hydrogen would be absorbed, and then the gas would pass on to a lime box, where the carbon dioxide would be absorbed, as already stated; however, whichever procedure is adopted as to the purification, a gas of the following approximate composition is obtained:—

Hydrogen	99·0
Carbon dioxide	nil
„ monoxide	·5
Sulphuretted hydrogen	trace
Nitrogen (by difference)	·5
	100·0

Secondary Chemical Reactions.—The fundamental chemical reactions, whereby hydrogen is produced by the use of water gas and steam alternately, in the presence of iron oxide, have now been given in considerable detail, and so far there does not appear any reason why the same iron ore should not be used indefinitely; however, there are two reasons which necessitate the replacement of the ore from time to time. The first reason for the deterioration of the ore is purely physical, while the second is partly chemical and partly physical. The physical reason for the gradual failure of the material is due to the fact that with constant use the ore tends to break up into smaller and smaller pieces, thus creating back pressure to the flow of water gas and steam; consequently a condition arises from this disintegration of the ore which necessitates its replacement.

When carbon monoxide is in contact with hot metallic iron, the following reaction slowly takes place :—

$$Fe + 6CO = FeC_3 + 3CO_2.$$

Such a condition arises in the Iron Contact process towards the end of the reducing stage, while during the oxidising stage the following reaction slowly takes place :—

$$3FeC_3 + 13H_2O = 9CO + Fe_3O_4 + 13H_2.$$

Thus, by the continued operation of the process, there tends to be an increasing amount of carbon monoxide in the resulting hydrogen. There are two methods whereby this difficulty can be dealt with : one is anticipatory, and consists in adding a volume of steam[1] to the water gas, prior to its passage over the iron oxide, equal to about one-half the carbon monoxide content of the gas. This, while slightly retarding the speed of reduction of the oxide, prevents the absorption of carbon by the metallic iron, formed during the reduction, and consequently allows of hydrogen of high purity being produced. The other method is intermittently employed and consists in occasionally passing air over the carbon-contaminated iron oxide, when the following reaction takes place :—

$$4FeC_3 + 15O_2 = 2Fe_2O_3 + 12CO_2,$$

thus allowing after reduction a purer hydrogen to be made. However, this process, known as "burning off," while undoubtedly improving the purity of the hydrogen subsequently produced, appears to hasten the disintegration of the oxide, contributing to the necessity for its ultimate replacement, owing to the high back pressure this physical condition produces.

[1] French patent 395132, 1908, Dellwick-Fleischer Wassergas Ges.

The minute quantities of sulphuretted hydrogen present in the crude hydrogen arise from two causes, the first of which is sulphur in the original ore, which, during the oxidising stage, produces sulphuretted hydrogen, while the other is due to the small quantities of sulphuretted hydrogen present in the purified water gas, which during the reducing stage are absorbed by the iron in the retorts as ferrous sulphide, which is subsequently decomposed during the oxidising stage, thus :—

$$FeS + H_2O = FeO + H_2S,$$
$$3FeO + H_2O = Fe_3O_4 + H_2.$$

With regard to the sulphuretted hydrogen, which is produced merely from the sulphur originally contained in the ore, this decreases with time; ore which when put in the retorts contained ·75 per cent. of sulphur, after a year in continuous use contained only 0·03 per cent.

Iron Contact Plant.—The fundamental and secondary chemical reactions involved in this process having been considered, there remains only the plant, and the actual fuel consumption per 1000 cubic feet of hydrogen to be described.

The Iron Contact plant is commercially manufactured in two distinct types :—

1. The Multi-Retort type.
2. The Single Retort type.

Fig. 10 shows a purely diagrammatic arrangement of a multi-retort generator. The retorts are externally heated by means of a gas producer incorporated in the retort bench. The even heating of the retorts is secured by the use of refractory baffles (not shown) and by the admission of air for the proper combustion of the producer gas at different points.

The retorts are arranged so that either blue water gas or steam can be passed through them by the operation of the valves A and B.

During the reducing stage the valves A and D are open, and B and C shut; thus the reducing gas passes through the oxide, and since in practice the whole of the carbon monoxide and hydrogen in the water gas is not used up in its passage through the retorts, it is passed

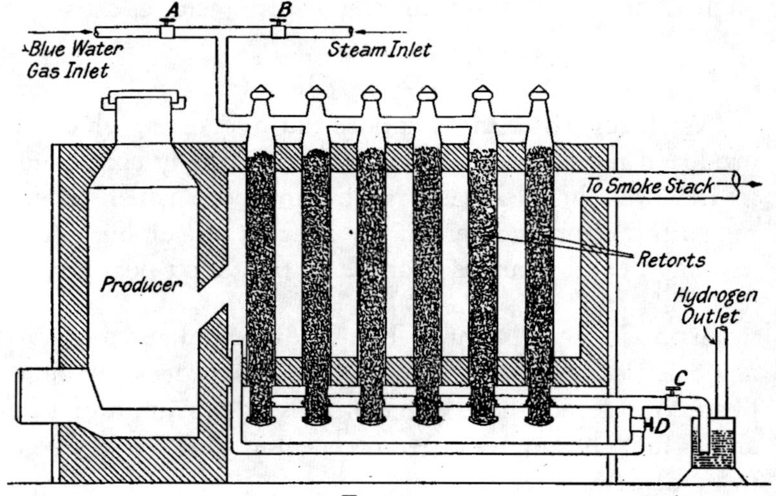

FIG. 10.

back outside of them, giving up its remaining heat, and consequently contributing to the external heating.

On the reducing stage being complete, the valves A and D are closed, and B and C opened; steam passes through the retorts, and hydrogen issues past the valve C to the water seal, and thence to scrubbers and purifiers, and finally to the gasholder.

When high purity hydrogen is required, on the reducing stage being complete, the valve A is first closed, and then the valve B turned on, allowing the hydrogen first made to pass, together with the residual gas, in the retorts via the valve D. When this purging has continued for about half a minute, C is opened and D closed, the hydrogen produced passing via the water seal ultimately to the gasholder.

Fig. 11 shows a diagram of a single retort plant

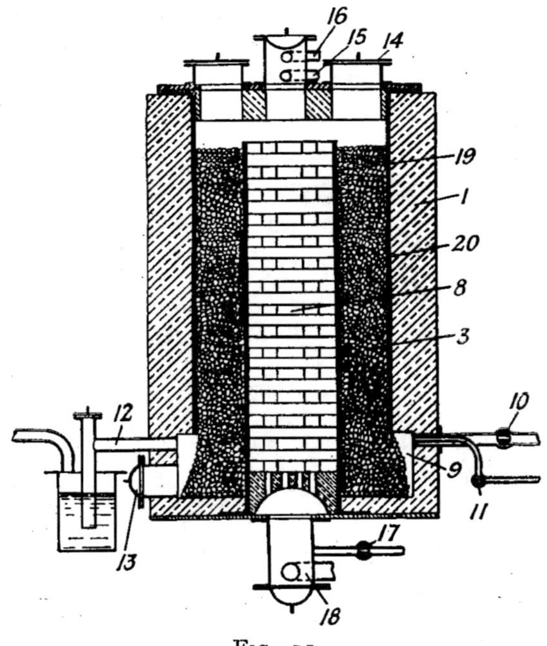

FIG. 11.

taken from Messerschmitt's specification, contained in English patent No. 18942, 1913.

This plant is circular in plan, and consists essentially of two cast-iron cylinders (19) and (20), the first of which is supported on its base and free to expand upwards, while the other is hung from a flange at its top, and is free to expand downwards. The annular space

(3) between the two cylinders is filled with suitable iron ore, while the circular space inside the smaller cylinder (19) is filled with a checker work of refractory brick (8). The plant is operated by first heating the refractory bricks (8) by means of water gas and air, admitted through pipes (15) and (16), the products of combustion going out to a chimney by the pipe (18). The heating of the checker work is communicated by conduction to the ore mass (3); when this is at a suitable temperature (about 750° C.) the gas supply (15) is shut and water gas enters by the pipe (10), passing up through the ore and reducing it in accordance with the equations already given. When the reducing gas reaches the top of the annular space (3) it mixes with air entering by the pipe (16) and the unoxidised portion (the amount of which varies, as has been shown in the graph, Fig. 9) burns, heating up the brick work, and finally passing away to the chimney by the pipe (18).

When reduction is complete (after about twenty minutes) pipes (16) and (10) are closed, and steam is admitted through the pipe (17), which passes upwards through the checker work (8) becoming superheated, and then down through the contact mass (3), where it is decomposed in accordance with the equations already given, producing hydrogen, which passes out by the pipe (12), through a water seal, and thence to a gas-holder. Where very pure hydrogen is required, a purging period can be introduced by adopting the following procedure :—

When reduction is complete, pipes (18) and (16) are closed, but pipe (10) is left open, and (12) still remains closed.

On the admission of steam by the pipe (17) hydrogen is generated in the reaction space (3), which, together with the residual water gas, is forced back into the water gas main (10), thus tending to increase the hydrogen content of the water gas in the gasholder.

After the lapse of sufficient time (about half a minute) pipe (12) is opened and (10) shut, the hydrogen subsequently produced passing via the water seal to the hydrogen holder. After the ore has originally been heated by means of water gas and air, admitted by pipes (15) and (16), the heat can be maintained entirely by the combustion of the unoxidised water gas, during the reducing stage, by the admission of air by the pipe (16).

"Burning off" can be accomplished by the admission of air by the pipe (11), the products passing out by the pipe (18). The top of the plant is fitted with four weighted valves, one of which is shown at (14). The Messerschmitt plant is not in commercial employment in this country, but it is considerably used both in Germany and in the United States, where the standard unit contains about 5 tons of iron ore, with a production of over 3000 cubic feet per hour.

Fuel Consumption.—In the multi-retort type of plant, the consumption of water gas is about 2·5 cubic feet per cubic foot of hydrogen produced, while in the single retort type, where the water gas is employed both for reduction and heating, the consumption is about 3·5 per cubic foot of hydrogen produced. In each type of plant, if the same kind of coke is used, both for the production of water gas and for all heating, including steam raising, both for the process and for its auxiliary machinery, such as blowers, feed pumps, etc., the hydrogen yield from each is about :—

6500-7000 cubic feet of hydrogen per ton of average soft coke.

Relative Advantages of the Multi- and Single Retort Plants.—While in fuel consumption there is little to choose between the two plants, there is undoubtedly less complication in the single retort plant than in the multi, owing to the fewer joints, etc., which are at high temperature.

Another advantage in the single retort type lies in the fact that fuel is consumed at two points only :—

1. For the production of steam, for the process and auxiliary machinery.
2. For the production of the necessary water gas.

In the multi-retort type, there is also fuel required for the supply of the producer, which heats the retort bench; however, this additional complication can be eliminated by heating the retorts externally by means of water gas, a procedure which is adopted in at least one commercial hydrogen plant.

With the gradual failure of the retorts themselves from their oxidation by the steam, the advantage again lies with the single retort type, as it is a simpler job to draw the cast-iron liners, and replace them, than it is to replace the individual retorts and make the various pipe joints.

To sum up, while in chemical efficiency there is little to choose between the two types, the advantage on the whole appears to lie with the single retort type, on account of its greater simplicity of repair.

The following patents with regard to this process are in existence :—

Oettli.	English	patent	16759.	1885.
Betou.	,,	,,	7518.	1887.
Lewes.	,,	,,	20752.	1890.
Hills & Lane.	,,	,,	10356.	1903.
Elworthy.	U.S.	,,	778182.	1904.
Vignon.	French	,,	373271.	1907.
Lane & Monteux.	,,	,,	386991.	1908.
Dellwick & Fleischer.	,,	,,	395132.	1908.
Lane.	English	,,	17591.	1909.
,,	,,	,,	11878.	1910.
Caro.	German	,,	249269.	1910.
Strache.	,,	,,	253705.	1910.
Messerschmitt.	U.S.	,,	971206.	1910.
,,	English	,,	12117.	1912.
,,	German	,,	263390.	1912.
,,	,,	,,	263391.	1912.
,,	,,	,,	268062.	1912.
Lane.	U.S.	,,	1028366.	1912.
,,	,,	,,	1040218.	1912.
Badische Anilin und Soda Fabrik.	French	,,	440780.	1912.
Messerschmitt.	,,	,,	461480.	1913.
,,	,,	,,	461623.	1913.
,,	,,	,,	461624.	1913.
,,	English	,,	18942.	1913.
Badische Anilin und Soda Fabrik.	French	,,	453077.	1913.
Badische Anilin und Soda Fabrik.	,,	,,	459918.	1913.

Lane.	U.S.	patent	1078686.	1913.
Berlin Anhaltische Maschinenbau.	English	,,	28390.	1913.
Pintoch.	French	,,	466739.	1913.
Berlin Anhaltische Maschinenbau.	English	,,	6155.	1914.

With Barium Sulphide.—In the previous process which was considered, steam was decomposed by means of spongy iron; in the present process, instead of iron, barium sulphide is used. If steam is passed over barium sulphide heated to a bright red heat, the following reaction takes place :—

$$BaS + 4H_2O = BaSO_4 + 4H_2.$$

The barium sulphate produced may be reduced by heating with coke to barium sulphide in accordance with the following equation :—

$$BaSO_4 + C = BaS + 4CO.$$

The barium sulphide can be employed for the generation of fresh hydrogen and the carbon monoxide can be used for supplying a portion of the heat which is required for the process.

The process is protected by French patent 361866, 1905, in the name of Lahousse.

A somewhat similar process to the Lahousse has been protected by French patent 447688, 1912, in the names of Teissier and Chaillaux. In this process barium sulphate is heated with manganous oxide, when the following reaction takes place :—

$$BaSO_4 + 4MnO = BaS + 4MnO_2.$$

The resulting mixture of barium sulphide and manganese dioxide is then raised to a white heat, when the following reaction takes place :—

$$BaS + 4MnO_2 = BaS + 4MnO + 2O_2.$$

When the reaction is complete, steam under pressure is passed over the mixture of barium sulphide and manganous oxide, with the production of hydrogen, in accordance with the following equation :—

$$BaS + MnO + 4H_2O = BaSO_4 + MnO + 4H_2.$$

The process is then ready to be started again. Whether it will have a considerable commercial application remains yet to be proved.

The Badische Catalytic Process.

Using a Catalytic Agent.—In the processes so far described for the production of hydrogen from steam, the steam has been decomposed by the action of some solid which itself undergoes a distinct chemical change requiring treatment to bring it back into a form in which it can be again used for the production of hydrogen. In the process about to be described the steam is decomposed by virtue of a catalytic agent which itself undergoes no permanent change.

This process, which is protected by patents (enumerated at the end of this note) by the Badische Anilin und Soda Fabrik Gesellschaft, consists of the following stages :—

First, Blue Water Gas is prepared in an ordinary producer and purified from suspended matter by means of a scrubber ; then into this clean water gas steam is introduced and the mixture passed over a catalytic material, where the following reaction takes place :—

$$\overbrace{H_2 + CO}^{\text{Water gas}} + H_2O = 2H_2 + CO_2.$$

Thus it is seen that the carbon monoxide contained in the blue gas is oxidised by the steam, which itself is decomposed with the production of hydrogen.

Now carbon dioxide is readily soluble in water, consequently the product of the reaction is passed under pressure through water, where it is absorbed, leaving a comparatively pure hydrogen.

Starting with blue water gas, which may be roughly taken as being composed of 50 per cent. hydrogen and 50 per cent. carbon monoxide, the composition of the gas, after the introduction of the steam and passage over the catalyst, is approximately as follows :—

	Per Cent. by Volume.
Hydrogen	65
Carbon dioxide	30
" monoxide	1·2-1·8
Nitrogen	2·5-4

The bulk of the carbon dioxide is absorbed by means of water, but if the hydrogen is required for aeronautical purposes, the gas is finally passed through either a caustic soda solution or over lime. Traces of carbon monoxide are removed by passing the gas under pressure through ammoniacal cuprous chloride solution. As a result of these final purifications a gas is obtained of approximately the following composition :—

	Per Cent. by Volume.
Hydrogen	97
Nitrogen	2·7
Carbon dioxide	—
" monoxide	·3

In practice it was stated that in commercial iron-contact plants the consumption of blue gas was from

2·3 to 3·5 cubic feet per cubic foot of hydrogen ultimately produced.

In the method which has just been described the consumption of blue gas is about 1·1 to 1·3 cubic feet per cubic foot of hydrogen, or assuming a consumption of 35 lb. of coke per 1000 cubic feet of water gas produced, the hydrogen yield is 49,000 to 58,000 cubic feet per ton of soft coke.

In the operation of this process, the blue water gas, together with a requisite amount of steam, is passed over the catalytic material at a temperature of 400° to 500° C. Since the oxidation of the carbon monoxide is exothermic, after the reaction chamber is heated to the temperature of 400° to 500° C., no more heat need be supplied from external sources.

The chemical composition of the catalyst appears to be somewhat variable, but, as in the case of the catalyst used in the fat-hardening industry, its physical condition effects the efficiency of the process. In the patents protecting this process a variety of methods are described for the preparation of the catalyst, but the following may be given as representative :—

"Dissolve a mixture of 40 parts by weight of ferric nitrate, 5 parts of nickel nitrate, and 5 parts of chromium nitrate, *all free from chlorine.* Precipitate with potassium carbonate, filter, wash, form into masses and dry."

The quantity of nickel can be varied, for example, between the limits of 10 parts and three parts of nickel nitrate.

This contact mass is used at a temperature of 400° to 500° C.

As is true of all catalysers, the above appears to be subject to "poisoning," the chief poisoners being chlorine, bromine, iodine, phosphorus, arsenic, boron, and silicon in some forms; hence in the preparation of the catalyser, as well as in the manufacture of the water gas, precautions must be taken to prevent the presence of these "poisons".

The mixture of blue water gas and steam is passed over the catalyst at approximately atmospheric pressure. On leaving the reaction chamber after passage through suitable regenerators, the gas is compressed to a pressure of 30 atmospheres (441 lb. per sq. inch) and then passes to the bottom of a high tower packed with flints, in which it meets a downward flow of water which absorbs the carbon dioxide, and also the sulphuretted hydrogen which is present in the gas to a very slight extent. The energy in the water leaving the tower is recovered in the form of power by letting it impinge on a Pelton wheel.

The removal of the 2 per cent. of carbon monoxide is accomplished in a similar tower; only in this case a solution of ammoniacal solution of cuprous chloride is used instead of water. Given an adequate size of tower and volume of the cuprous chloride solution, the pressure at which the gas is introduced into the tower may be as low as 30 atmospheres (441 lb. per sq. inch); however, where the gas is to be used for making synthetic ammonia it is usual to compress it to 200 atmospheres (2940 lb. per sq. inch) before passing it through the carbon monoxide absorbing tower. The use of this high pressure is ultimately necessary in the ammonia process and it reduces the size of the tower which has to be employed.

The cuprous chloride solution, after leaving the absorption tower, is passed through a small vessel, in which it gives up its carbon monoxide. The gas evolved from the solution is passed through water in order to prevent any ammonia loss.

The advantages of this process over the Iron Contact process are :—

1. It is continuous in operation.

2. It is more economical.

3. The whole of the sulphur compounds in the blue gas are converted into sulphuretted hydrogen and are completely absorbed by the high pressure water scrubbing.

The disadvantages as compared with the Iron Contact process are :—

1. Greater complexity of operation.

2. For aeronautical purposes the percentage of nitrogen is high.

Description of Plant.—The diagram (Fig. 12) shows the method of operation of this process. Steam enters by the pipe A and mixes with blue water gas entering by the pipe B, the speed of flow of each being indicated on separate gauges as shown. The mixture of steam and gas passes through the regenerator or superheater C and flows, as indicated by the arrows, over refractory tubes, through which the hot products of the reaction are flowing in the reverse direction. The heated mixed gases flow via the pipe F into the generator and, increasing in temperature, pass through the catalytic material, where reaction takes place with the evolution of heat. They then flow as indicated by the arrows back through the regenerator, parting with their heat to the incoming mixture of blue water gas and steam.

Thermo-couples are placed in the contact mass so that its temperature may be controlled by increasing or reducing the quantity of steam in the incoming gaseous mixture.

The whole apparatus is very effectively lagged to reduce the heat losses to a minimum.

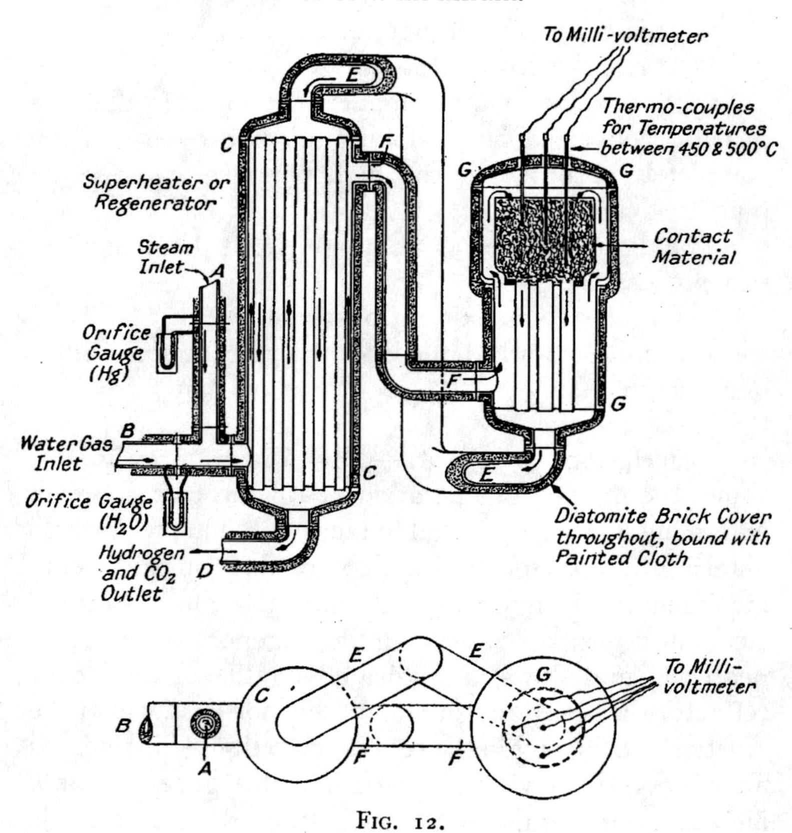

FIG. 12.

The following patents on this process by the Badische Analin and Soda Fabrik are in existence :—

English	patent	27117.	1912.
,,	,,	27963.	1913.
French	,,	459918.	1913.

The following patents relating to the general chemical reaction in this process have been taken out :—

Tessie du Motay.	U.S.	patent	229338.	1880.
,, ,, ,,	,,	,,	229339.	1880.
,, ,, ,,	,,	,,	229340.	1880.
Pullman & Elworthy.	English	,,	22340.	1891.
Elworthy.	French	,,	355324.	1905.
Ellis & Eldred.	U.S.	,,	854157.	1907.
Chem. Fabrik Greisheim Elektron.	British	,,	2523.	1909.
Naber & Muller.	German	,,	237283.	1910.

Using Lime.—If carbon monoxide together with steam is passed over lime at a temperature of about 500° C., the monoxide is absorbed with the formation of calcium carbonate, and hydrogen is evolved in accordance with the following equation :—

$$CaO + H_2O + CO = CaCO_3 + H_2.$$

Investigation of the above reaction by Levi & Piva[1] indicates that the chemical change takes place in two stages, in the first of which calcium formate is produced, while in the second it is decomposed with the evolution of hydrogen and carbon monoxide as is shown in the following equations :—

$$(1)\ CaO + H_2O + 2CO = Ca(COOH)_2,$$
$$(2)\ Ca(COOH)_2 = CaCO_3 + CO + H_2.$$

It can, however, be seen from these equations that whatever volume of carbon monoxide is permanently absorbed, an equal volume of hydrogen is evolved.

Now, since blue water gas is, roughly speaking, half hydrogen and half carbon monoxide, by passing it over

[1] "Journ. Soc. Chem. Ind.," 1914, 310.

lime under the conditions stated above, a gas equal in volume to the water gas, but wholly composed of hydrogen, is produced.

In the commercial operation of this process, the lime is contained in a tower, which is initially heated to a temperature of about 500° C., but since the absorption of the carbon monoxide is exothermic, after the process has started, no further heating is required.

When the lime has become sluggish in its action, by the formation of a crust of calcium carbonate, the blue gas is diverted through a similar tower, while the contents of the original tower are heated *in situ* to a temperature sufficiently great to decompose the calcium carbonate, and thus the tower is again ready for use.

This process is protected by the following patents :—

Chem. Fabrik Greisheim Elektron.	British	patent	2523.	1909.
Dieffenbach & Moldenhauer.	,,	,,	8734.	1910.
Ellenberger.	U.S.	,,	989955.	1912.
Chem. Fabrik Greisheim Elektron.	British	,,	13049.	1912.

(4) Miscellaneous Methods of Making Hydrogen.

THE CARBONIUM GESELLSCHAFT PROCESS.

From Acetylene.—If acetylene is compressed and then subjected to an electric spark it undergoes dissociation into its elements.

Acetylene can be most easily generated from the action of water on calcium carbide, thus :—

$$CaC_2 + 2H_2O = C_2H_2 + Ca(OH)_2.$$

The acetylene produced is then compressed in very strong cylinders and subjected to an electric spark, when the following reaction takes place :—

$$C_2H_2 = C_2 + H_2.$$

If the acetylene is produced from calcium carbide, approximately 178 lb. of calcium carbide and 100 lb. of water are theoretically required to produce 1000 cubic feet of hydrogen at 40° F. and 30 inches barometer, while, at the same time, 39 lb. of carbon in the form of lamp-black is produced.

This process is employed by the Carbonium Gesellschaft of Frederickshaven for the inflation of airships, while the carbon produced is sold and is used in making printers' ink. As used by this company, the gas is compressed to about 2 atmospheres (29·4 lb. per sq. inch) prior to sparking.

The following patent, relative to this process, is in existence :—

Bosch. German patent 268291. 1911.

The decomposition of acetylene may be obtained by heating ; thus, if acetylene derived from calcium carbide or some other source is passed through a tube heated to about 500° C. it decomposes, in accordance with the following equation, with the evolution of heat :—

$$C_2H_2 = C_2 + H_2.$$

Such is the quantity of heat liberated that after the temperature of the tube has been raised until decomposition of the acetylene begins no further external heat is required.

The carbon produced may be chiefly removed by filtering the gas, while the residue which still remains may be removed by scrubbing with water.

This process is protected by the following patents :—

Picet.	French patent		421838.	1910.
,,			421839.	—
	English	,,	24256.	1910.
	German	,,	255733.	1912.

From Hydrocarbon Oils.—While the decomposition of acetylene is attended with the evolution of heat, most other hydrocarbon gases absorb heat when they decompose into their constituents; consequently, to produce hydrogen from other hydrocarbon gas or volatilised hydrocarbon oils, it is necessary to supply heat during the process.

The necessary heat may be supplied by passing the hydrocarbon gas or vaporised oil through a tube of refractory material which is externally heated, or the ingenious Rincker-Wolter method may be used. In this process the rough principle is to use a generator similar to a "blue-gas" generator filled with coke. By means of an air blast the temperature of the coke is raised to about 1200° C., then, when this temperature has been reached, the air supply is stopped and crude oil is blown in at the bottom of the hot coke.

The oil is immediately volatilised, and passes by expansion up through the hot coke, during which process it is decomposed into hydrogen and carbon, the latter to a large extent attaching itself to the coke and becoming a source of fuel. When the temperature has fallen too low to effect a complete decomposition of the crude oil the injection is stopped and the temperature of the coke again raised by means of the air blast. The gas produced by this process is stated to have the following composition :—

	Per Cent. by Volume.
Hydrogen	96·0
Nitrogen	1·3
Carbon monoxide	2·7

The cost of hydrogen made by this process must depend almost entirely on the price of crude oil; it is stated that, with crude oil at twopence a gallon, hydrogen can be produced for about seven shillings a thousand cubic feet.[1]

The following patents deal with this or somewhat similar processes:—

Geisenberger.	French patent	361462.	1905.
Rincker & Wolter.	,, ,,	391867.	1908.
,, ,,	,, ,,	391868.	1908.
Berlin Anhaltische Maschinenbau A. G.	German ,,	267944.	1913.
	French ,,	466040.	1913.
	English ,,	2054.	1914.
C. Ellis.	U.S. ,,	1092903.	1914.

From Starch.—When yeast, which is a living organism, is introduced into a solution containing sugar, fermentation results with the production of alcohol and carbon dioxide, which may be expressed in an equation as follows:—

$$C_{12}H_{22}O_{11} + H_2O = 4C_2H_6O + 4CO_2.$$

An analogous process to the above is employed commercially for the production of acetone and butyl alcohol.

When what is known as the Fernbach bacillus is introduced into starch jelly, $n(C_6H_{10}O_5)$, acetone, $(CH_3)_2CO$,

[1] Ellis, "The Hydrogenation of Oils" (Constable).

and butyl alcohol, $CH_3CH_2CH_2CH_2OH$, are produced; at the same time there is an evolution of gas which is chiefly hydrogen and carbon dioxide, but it also contains a little nitrogen.

As there is a great demand for acetone in certain localities, large quantities of hydrogen in this impure form are being produced as a by-product. If the carbon dioxide is absorbed by passing the gas under pressure through water (Bedford method), a gas is produced of about the following composition :—

Hydrogen	94·0
Nitrogen	6·0

The above is not a process for the production of hydrogen, but the hydrogen produced may be frequently usefully employed if there is a local demand for it.

CHAPTER IV.

THE MANUFACTURE OF HYDROGEN.

CHEMICO-PHYSICAL METHODS.

Linde-Frank-Caro Process.—The most important method of producing hydrogen, in which chemical and physical methods are employed, is one in which the chemical process results in the production of blue water gas, and the physical in the separation of the chemical compounds (chiefly carbon monoxide) from the hydrogen by liquefaction.

THE SEPARATION OF HYDROGEN FROM BLUE WATER GAS.

The separation of mixed gases by liquefaction is a subject of very great complexity and one into the intricacies of which it is not intended to go in this work, but for further information the attention of the reader is directed to the two following books :—

"The Mechanical Production of Cold," by J. A. Ewing. (Cambridge University Press.)

"Liquid Air, Oxygen, Nitrogen," by G. Claude. (J. & A. Churchill.)

All gases are capable of being liquefied, but in the case of hydrogen and helium[1] the difficulties are so great

[1] This gas, which was the last to resist liquefaction, was first liquefied on 10th July, 1908, by Professor Kamerlingh-Onnes.

that it is only by means of the highest technical skill and very costly apparatus that this can be accomplished.

Originally it was considered that to obtain a gas in the liquid state the sole necessity was pressure; however, all gases possess a physical property known as *critical temperature*.[1] The critical temperature of a gas is that temperature above which the gas cannot be liquefied, however great the pressure to which it is subjected.

Prior to the realisation of the existence of the critical temperature, chemists and physicists subjected various gases to enormous pressures in the hope of causing them to liquefy, and, though they failed, it is interesting to observe from the accounts of their experiments that the compressed gas attained a density greater than the same gas in the liquid state at atmospheric pressure.

Besides critical temperature, another term requires definition, that is, *critical pressure*, which is the pressure which must be exerted on a gas cooled to its critical temperature to produce liquefaction.

The following table of critical temperatures and pressures of the constituents of blue water gas is interesting:—

Gas.	Critical Temperature.	Critical Pressure.
Hydrogen . . .	− 234·0° C.	294 lb. per sq. in.
Carbon monoxide .	− 136·0	492 ,, ,,
,, dioxide . .	+ 30·92	1131 ,, ,,
Nitrogen . . .	− 146·0	485 ,, ,,
Methane . . .	− 82·0	820 ,, ,,
Sulphuretted hydrogen	+ 100·0	1304 ,, ,,
Oxygen . . .	− 118·0	735 ,, ,,

[1] Discovered by Andrews, 1863.

From this table it is seen that the critical temperature of hydrogen is 88° C. below that of its nearest associate, nitrogen; consequently, if the blue water gas were cooled to – 146° C. while subjected to a pressure of somewhere about 500 lb. per square inch, the whole of the gas, with the exception of the hydrogen, would liquefy; therefore, separation of a liquid from a gas being a simple matter, the problem of the production of hydrogen from blue water gas would be solved.

If a gas is cooled below its critical temperature the pressure which has to be applied to produce liquefaction is much reduced. Now, since the boiling point of a liquid and the condensing point of a vapour under the same pressure are the same temperature, the boiling points of the various gases contained in blue water gas can be studied with advantage.

BOILING POINTS OF SOME LIQUID GASES AT ATMOSPHERIC PRESSURE.

Gas.	Boiling Point.
Hydrogen	– 253·0° C.
Carbon monoxide	– 190·0
" dioxide	– 80·0
Nitrogen	– 195·5
Methane	– 164·7
Sulphuretted hydrogen	– 61·6
Oxygen	– 182·5

Therefore, it can be seen that, if blue water gas were cooled at atmospheric pressure to a temperature below – 195·5° C., the whole of the constituents of the gas, other than hydrogen, would be liquefied, and consequently an easy separation could be made.

To summarise, the liquefaction of the constituents of

blue water gas, other than hydrogen, can be accomplished either by a moderate degree of cooling and the application of pressure, or by intense cooling and no application of pressure.

PRODUCTION OF HYDROGEN FROM WATER GAS BY THE LINDE PROCESS.

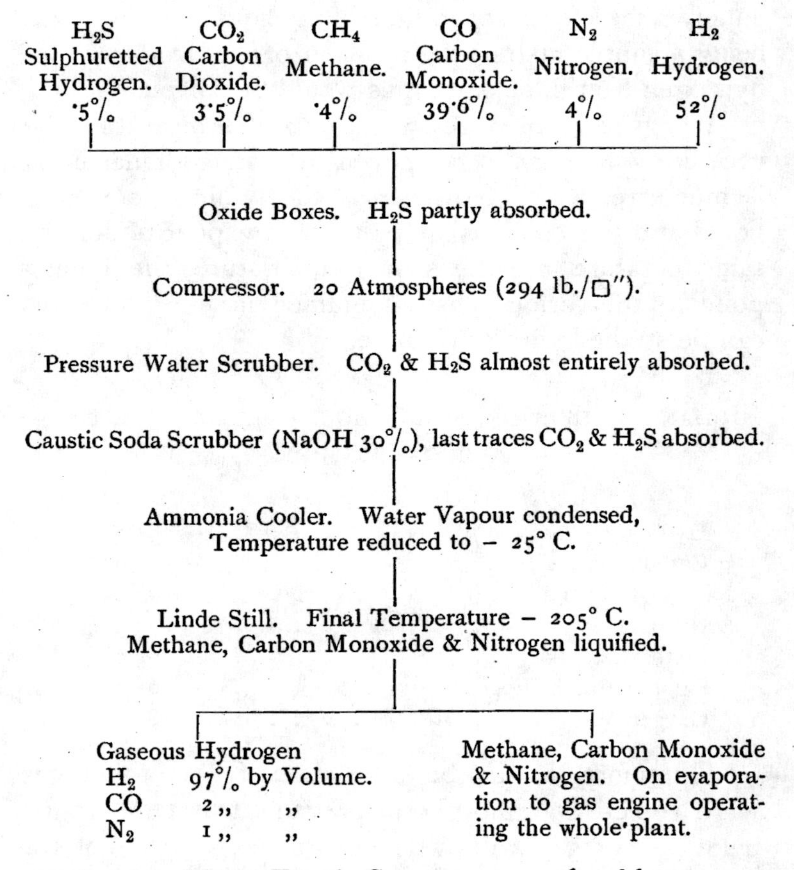

In the Linde-Frank-Caro process the blue water gas is compressed to 20 atmospheres, and under pressure it is passed through water, which removes practically the whole of the carbon dioxide and sulphuretted hydrogen. It is then passed through tubes containing caustic soda, which removes the remaining traces of carbon dioxide, sulphuretted hydrogen, and water.

The gas thus purified from these constituents now passes to the separator proper ; the reason for this preliminary removal of some of the constituents of the blue water gas is due to the fact that, in the separation of the carbon monoxide and nitrogen, such low temperatures have to be reached that the water, sulphuretted hydrogen, and carbon dioxide would be in the solid state, and would, therefore, tend to block up the pipes of the apparatus.

The apparatus operates in the following manner, which will be more readily understood by consulting the diagram (Fig. 13):—

The purified water gas passes down the tube A, through coils in the vessel B, which is filled with liquid carbon monoxide boiling at atmospheric pressure (– 190° C.). Now, since the water gas is under pressure and is passing through coils cooled to its temperature of liquefaction at atmospheric pressure, the bulk of it liquefies (theoretically more gas should be liquefied in the tubes than is evaporated outside them).

The gas, containing a considerable amount of liquid saturated with hydrogen, passes into the vessel C, which is surrounded by liquid nitrogen boiling under reduced pressure giving a temperature of – 205° C. Here the remainder of the carbon monoxide and the nitrogen originally contained in the gas liquefies and hydrogen of approximately the following composition passes up the tube E :—

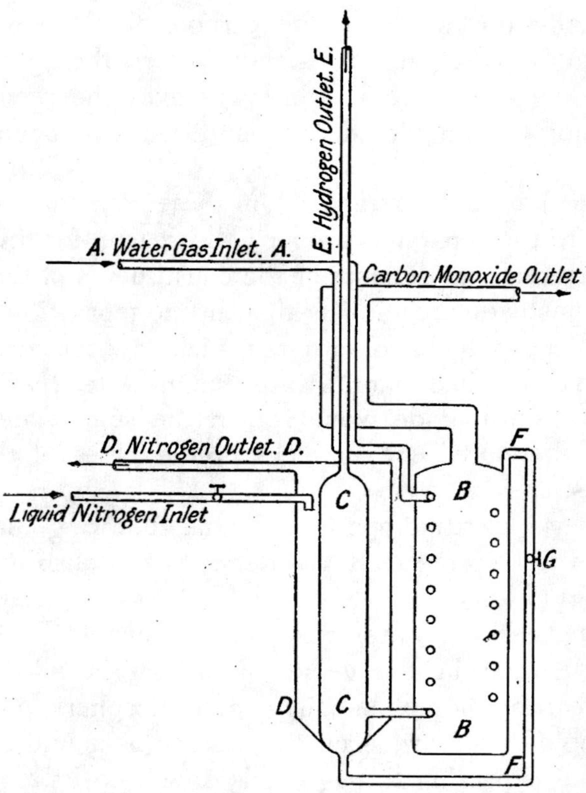

FIG. 13.—Diagram showing Linde-Frank-Caro Process.

	Per Cent. by Volume.
[1] Hydrogen	97·0
Nitrogen	1·0
Carbon monoxide	2·0
Sulphuretted hydrogen	nil.
Organic sulphur compounds	—

When the gas is required to be of high purity it is

[1] Messrs. Ardol of Selby, Yorks, who employ this process, kindly supplied the author with this information.

subsequently passed over calcium carbide or soda lime, the reactions of which processes will be dealt with later.

During the operation of the process liquid carbon monoxide and some liquid nitrogen collects in C. Now this liquid gas is under pressure and can therefore be run back through the tube F via the cock G into the vessel B; but B is at atmospheric pressure, consequently some of the liquid gas passing through G will be volatilised, with consequent fall in temperature of the remainder.

The liquid nitrogen used in the vessel D is produced in a special Linde machine from the atmosphere.

The vapour of carbon monoxide, with a little nitrogen and hydrogen, from the vessel B is used to cool the incoming purified water gas, as is shown in the diagram. This method of using the cold separated gases for cooling the gas going into the apparatus is termed "Cooling by counter-current heat exchangers," and it may be regarded as the essence of efficiency in all low temperature gas separation.

The consumption of power in this process is theoretically very small, as much carbon monoxide should be liquefied in the coil in the vessel B as is volatilised outside it (this is theoretically true when the pressure of the gas passing through the coil is atmospheric). However, in practice, the necessity for power consumption arises from the fact that liquid nitrogen must be continuously supplied to the vessel D in order to prevent the temperature of the plant rising from external infiltration of heat, which takes place in spite of the most effective lagging.

In practice, the power obtained from using the separated carbon monoxide as a fuel is sufficient to run all the machines necessary for the operation of a plant producing 3500 cubic feet of hydrogen or more per hour.

Thus, to very roughly indicate the cost of operation of this process, neglecting all depreciation, etc., it may be said that, on a plant of the size mentioned above, unit volume of blue water gas yields ·4 volume of hydrogen of about 97 per cent. purity, or, on the basis of a coke consumption of 35 lb. per 1000 cubic feet of water gas, the hydrogen yield is 25,500 cubic feet per ton of coke.

Purification of Hydrogen.—Where very pure hydrogen is required it is necessary to employ chemical methods to remove the 3 per cent. of impurity, which may be done by passing the gas through heated soda lime, where the carbon monoxide is absorbed in accordance with the following equation :—

$$2NaOH + CO = Na_2CO_3 + H_2,$$

or, on the other hand, it may be passed through heated calcium carbide (over 300° C.), which possesses the advantage of not only removing the carbon monoxide but also the nitrogen. The reactions taking place are indicated in the following equations :—

$$CaC_2 + CO = CaO + 3C,$$
$$CaC_2 + N_2 = CaCN_2 + C.$$

The following is given as an analysis of the gas purified by means of soda lime :—

	Per Cent. by Volume.
Hydrogen	99·2-99·4
Carbon monoxide	nil.
Nitrogen	0·8-0·6

The following patents are in existence for the production of hydrogen by liquefaction methods from blue water gas :—

Elworthy.	French patent		355324.	1905.
Frank.	English	,,	26928.	1906.
Claude.	French	,,	375991.	1906.
Ges-fur Linde's Eis-maschinen A.G.	,,	,,	417983.	1911.
C. von Linde.	U.S.	,,	1020102.	1912.
			1020103.	1912.
			1027862.	1912.
			1027863.	1912.
Chemical purification—				
Frank.	French	,,	371814.	1906.

Diffusion.—The separation of hydrogen from the other constituents of blue water gas has been proposed, employing diffusion for the purpose. Graham expressed the law of diffusion of gases as :—

"The relative velocities of diffusion of any two gases are inversely as the square roots of their densities."

That is to say, if a mixture of two gases of different densities is passed through a porous tube, e.g. unglazed porcelain, in a given time, more of the lighter gas would have passed through the walls of the tube than of the heavier, or, to take a concrete example, suppose the mixture of gases was one composed of equal parts by volume of hydrogen and oxygen, then, since their densities are as 1 to 16, and since, therefore, the roots of their densities are as 1 to 4, in a given time four times as much hydrogen would diffuse through the medium as oxygen.

The densities and the square roots of the densities of the constituents of blue water gas are given below ;—

	Density.	$\sqrt{D}$.
Hydrogen	1	1
Carbon monoxide . .	14	3·7
,, dioxide . . .	22	4·7
Nitrogen	14	3·7
Methane	8	2·8
Sulphuretted hydrogen .	17	4·1
Oxygen	16	4·0

From which it will be seen that, if blue gas were passed continuously through a porous tube, the gas diffusing through the tube would contain more hydrogen than the blue gas originally contained. Of course, in the successful operation of a diffusion separation it is necessary to remove the gas which diffuses through the porous medium as well as the residue which is left undiffused. The former may be done by maintaining a constant pressure by means of a suction pump, while the latter can be done by regulating the speed of flow through the diffusion tube. It is, of course, essential that the undiffused gas must be removed from contact with the porous medium after a certain time, as it is only a matter of time before the whole of the gas will diffuse through the medium, and thus destroy the work of separation.

The Diffusion Medium.

The selection of the diffusion material is a subject of considerable difficulty; if the porosity of the material is too great no diffusion takes place, but the gas passes through the material without any appreciable separation taking place. Thus, if a mixture of hydrogen and oxygen is passed through a fine capillary tube, the gas issuing will be found to be of the same composition as the original gas.

It is interesting to note in this connection that, if pure hydrogen were first passed through the tube and then pure oxygen, in a given time more hydrogen by volume would pass through the tube than oxygen. This differential rate of flow through tubes is called "Transpiration".

If the porosity of the material is insufficient, the time required to effect separation is unduly long. It may, in this connection, be mentioned that it has from time to time been suggested that by means of diffusion it would be possible to separate a mixture of gases of different densities without the consumption of power.[1] However, in practice this has not been found to be the case, as, in order to obtain a reasonable speed of separation, a difference of pressure between the two sides of the diffusion material has to be maintained.

Jouve and Gautier have employed a diffusion method in order to separate hydrogen from blue water gas, and it is stated that, by a single passage through a porous partition, the percentage of carbon monoxide in the gas passing through the medium was reduced from 45 per cent. in the original gas to 8 per cent. Whether this process has been employed on a commercial scale is not known to the author, nor has he any knowledge as to the amount of power required to obtain a definite volume of hydrogen practically free from carbon monoxide.

The following patents, in which diffusion has been

[1] It is theoretically impossible to separate a mixture of two gases without the consumption of power, but the theoretical requirements are almost negligible.

employed for separating mixed gases, have been taken out :—

Jouve & Gautier.	French patent	372045.	1908
Hoofnagle.	U.S. „	1056026.	1913

Separation by Centrifugal Force.—When a mass is compelled to move in a circular course a force acts on it which is a function of its mass, linear velocity, and the radius of curvature of its path, which may be expressed as—

$$\text{Centrifugal force} = \frac{m \cdot v^2}{R}$$

where m = mass of the body,
v = its linear velocity,
R = the radius of curvature of its path.

Therefore, since a greater force is acting on the heavier of two particles moving on the same course with the same velocity, the heavier particle will tend to move outward from its centre of rotation to a greater extent than the lighter. This principle of centrifugal force is employed industrially for many purposes, such as the separation of cream from milk, water from solid bodies, honey from the comb, etc., and it has been suggested that it might be used to separate hydrogen from blue water gas. However, though a certain amount of work has been done on this problem by Elworthy[1] and Mazza,[2] as far as the author knows no satisfactory results have been obtained.

The special physical questions involved in the separation of gases of different densities by means of a centrifugal machine have been considered theoretically

[1] Elworthy, English patent, 1058. 1906.
[2] Mazza, English patent, 7421. 1906.

by a number of physicists, whose conclusions are that very high velocities must be given to the gas to obtain any appreciable separation; it has been shown that, if a drum 3 feet in diameter and one foot long filled with a mixture at 15° C., containing 80 per cent. of hydrogen and 20 per cent. of air, is rotated at 20,000 revolutions per minute, a condition of dynamical equilibrium will arise when the peripheral gas and the axial gas will have the following composition :—

	Axial Gas.	Peripheral Gas.
Hydrogen	97·8	66·1
Air	2·2	33·9

Since the density of air and that of carbon monoxide are almost the same (14·4 and 14·0) almost identical theoretical results could be obtained by giving a similar rotary motion to a mixture of 80 per cent. hydrogen and 20 per cent. carbon monoxide. However, the enormous speed of rotation and a practical method of removing the axial and peripheral gases makes this question one of the greatest technical difficulty, and it may well be that the power consumption to produce a given volume of hydrogen from blue water gas may be greater than that required to produce an equal volume of hydrogen by electrolysis.

CHAPTER V.

THE MANUFACTURE OF HYDROGEN.

Physical Methods.

Electrolysis.—When an electric current passes through a solid conductor a magnetic field is created round the conductor and the conductor is heated by the passage of the current, both of which effects bear a definite relationship to the magnitude of the current passing. Some liquids are also conductors of electricity, e.g. mercury; the passage of a current through such a conductor produces results identical with those produced in solid conductors. Other liquids are also conductors, but, besides the passage of the current creating a magnetic field and a heating effect, a portion of the liquid is split up into two parts which may each be a chemical element, or one or either may be a chemical group.

Thus, if two platinum plates are placed as shown in the diagram, one plate being connected to the positive pole of the battery and the other to the negative, then, if a strong aqueous solution of hydrochloric acid is put in the vessel containing the plates, decomposition of the liquid will take place: hydrogen will be given off at the negative plate or cathode, and chlorine at the positive or anode.

If the solution of hydrochloric acid is replaced by one of caustic soda the caustic soda is split up by the current into oxygen, which is liberated at the anode, and metallic sodium which is deposited on the cathode; but since metallic sodium cannot exist in contact with water, the following reaction takes place at the cathode :—

$$2Na + 2H_2O = 2NaOH + H_2.$$

Thus, by a secondary reaction, hydrogen is liberated at the cathode, or, in other words, water is split into its constituents, while the caustic soda is reformed.

Now, let the caustic soda solution be replaced by an aqueous solution of sulphuric acid. In this case hydrogen will be liberated at the cathode and the group SO_4 at the anode, but the group SO_4 cannot exist in contact with water, as the following reaction takes place :—

$$2SO_4 + 2H_2O = 2H_2SO_4 + O_2.$$

Thus, by a secondary reaction, oxygen is liberated at the anode, or, in other words, water is split into its constituents while the sulphuric acid is reformed.

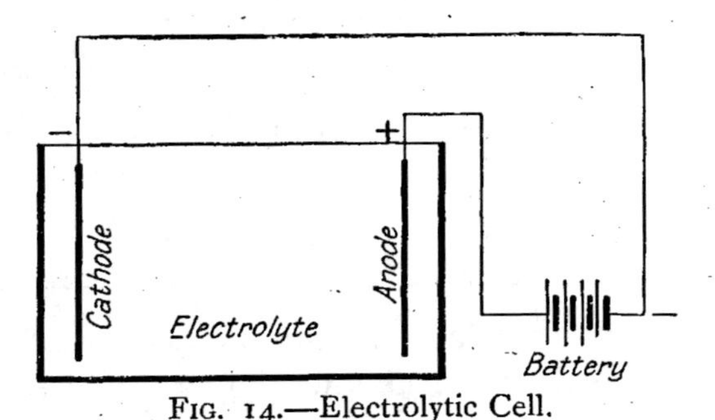

FIG. 14.—Electrolytic Cell.

Liquids which, under the influence of the electric current, behave in the manner of the above are termed "Electrolytes," and the process whereby they are split up is called "Electrolysis".

The laws relating to this decomposition of liquids by the electric current were enunciated by Faraday as follows:—

1. The quantity of an electrolyte decomposed is proportional to the quantity of electricity which passes.

2. The mass of any substance liberated by a given quantity of electricity is proportional to the chemical equivalent weight of the substance.

By the *chemical equivalent weight* of a substance is meant in the case of elements, the figure which is obtained by dividing its atomic weight by its valency, while in the case of compounds, it is the molecular weight divided by the valency of the compound. However, many elements have more than one valency, therefore they have more than one chemical equivalent weight, as can be seen from the following table:—

Element.	Atomic Weight.	Valency.	Chemical Equivalent Weight, $\frac{A. T.}{V.}$.
Hydrogen . .	1	1	1
Oxygen . .	16	2	8
Gold . .	197	3 or 1	65·6 or 197
Tin . . .	118	4 „ 2	29·5 „ 59
Phosphorus .	31	5 „ 3	6·02 „ 10·03
Tungsten . .	184	6 „ 4	30·6 „ 46·0

From Faraday's laws it can be seen that, if the weight of any substance liberated by a definite current in a definite time is known, the theoretical weight of any substance which should be liberated by a definite current in a definite time can be calculated, if the chemical equivalent weight of this substance is known. Very careful experiments have been made with regard to the amount of silver deposited by a current of one ampere flowing for one second (one coulomb); this current deposits

$$\cdot 001118 \text{ gram of silver}$$

from an aqueous solution of a silver salt.

Now the atomic weight of silver is

$$107 \cdot 94$$

and its valency is unity, therefore its chemical equivalent weight is

$$107 \cdot 94,$$

but the atomic weight of hydrogen is

$$1 \cdot 0$$

and its valency is unity, therefore its chemical equivalent weight is

$$1 \cdot 0,$$

therefore it follows from Faraday's second law that $\frac{\cdot 001118}{107 \cdot 94} = \cdot 000010357$ gram of hydrogen will be liberated by one ampere flowing for one second, or the mass of hydrogen liberated by any current in any time may be expressed as

$$1 \cdot 0357 \times 10^{-5} \mathrm{A}t$$

where A is the current in amperes and t the time it flows in seconds; which is equivalent to saying that, at 0° C. and 760 mm. barometric pressure (29·92 inches), one ampere-hour will liberate

$$\cdot 0147 \text{ cubic foot of hydrogen.}$$

So far the relationship between current and volume of hydrogen which would be produced theoretically has been considered; it now remains to determine the relationship between power and the volume of hydrogen which should be theoretically liberated. To refer to the diagram, it will be at once appreciated that, to get the current to flow through the electrolyte requires an electrical pressure, or, in other words, there will be found to be a voltage drop between the anode and cathode.

This voltage drop is due to two types of resistance, one of which is identical to the resistance of any conductor and is dependent on the cross-sectional area of the path of flow of the current and on the length of the path, i.e. the distance between the plates. The other resistance is one that is due to a condition analogous to the back E.M.F. of an electric motor. Assume that electrolysis has been taking place in the diagrammatic cell and that the battery has been removed; if a voltameter is then placed between the anode and cathode it will be found that there is a difference of potential between the two plates and that the direction of this electromotive force is the reverse of that of the current which was supplied in the first instance by the battery. This resistance is called the back E.M.F. of the cell, or the polarisation resistance. While the first type of resistance can be practically eliminated by placing the plates close together, the second is not a function of the cell design, but a constant of the electrolyte in the cell; therefore, to obtain electrolysis in a cell it is necessary that the current must have a certain theoretical potential to overcome the polorisation resistance of the electrolyte.

The minimum voltage to produce continuous electrolysis in a cell whose resistance other than that due to polarisation is negligible is given below for various aqueous solutions of bases, acids and salts containing their chemical equivalent weight in grams per litre ; with considerable variation in the degree of concentration of the solution it has been found that those solutions given below whose minimum voltage is about 1·7 require no appreciable variation in pressure to produce continuous electrolysis :—

Solution of	Minimum Voltage for Continuous Electrolysis.
Zinc sulphate	2·35 volts[1]
Cadmium sulphate	2·03 ,, [1]
,, nitrate	1·98 ,, [1]
Zinc bromide	1·80 ,, [1]
Cadmium chloride	1·78 ,, [1]
Orthophosphoric acid	1·70 ,, [1]
Nitric acid	1·69 ,, [1]
Caustic soda	1·69 ,,
,, potash	1·67 ,,
Lead nitrate	1·52 ,, [1]
Hydrochloric acid	1·31 ,, [1]
Silver nitrate	·70 ,, [1]

Now it has been previously deduced from Faraday's laws that a current of one ampere for one hour should produce ·0147 cubic foot of hydrogen (at 0° C. and 760 mm. pressure), but if a solution of caustic soda was used the current would have had to be supplied at 1·69 volts, therefore 1 × 1·69 watt-hour produces ·0147 cubic foot of hydrogen, or

$$1000 \text{ watt-hours produce } \frac{\cdot 0147 \times 1000}{1\cdot 69} = 8\cdot 7 \text{ cubic feet.}$$

But, at the same time as the hydrogen is liberated at the cathode, oxygen is being liberated at the anode, and since from Faraday's laws the volume of oxygen is one half of that of the hydrogen, on the electrolysis of a

[1] Determined by Le Blanc.

solution of caustic soda 1 kilowatt-hour (B.T.U) theoretically produces

8·7 cubic feet of hydrogen at 0° C. 760 mm. (29·92").
and 4·4 " " oxygen " "

The theory of electrolysis having been considered, it remains to describe some of the more important applications of this phenomenon for the production of hydrogen and oxygen.

To refer again to the diagrammatic cell, if the distance between the anode and cathode is great the resistance of the cell is high, and consequently the production of hydrogen is much below the theoretical, but if, on the other hand, the distance between the two plates is small, the gases liberated are each contaminated with the other, hence the design of a cell for the commercial production of oxygen and hydrogen has of necessity to be a compromise between these extremes.

A large number of commercial cells put the anode and cathode comparatively close together, but, in order to obtain reasonably high purity in the gaseous products, a porous partition is placed between the electrodes: this, like increasing the distance between the plates, creates a certain amount of resistance, but it has one advantage of the latter procedure in that it makes for compactness, which is very desirable in any plant and particularly so in the case of electrolytic ones, as one of the greatest objections to their use is the floor space which they occupy.

A glance at the list of patents at the end of the chapter will show what an amount of ingenuity has been expended in the design of electrolytic plant for the production of oxygen and hydrogen. On account of this multiplicity of different cells it is intended merely to describe the following, which are representative types :—

1. Filter press type.
2. Tank type.
3. Non-porous non-conducting partition type.
4. Metal partition type.

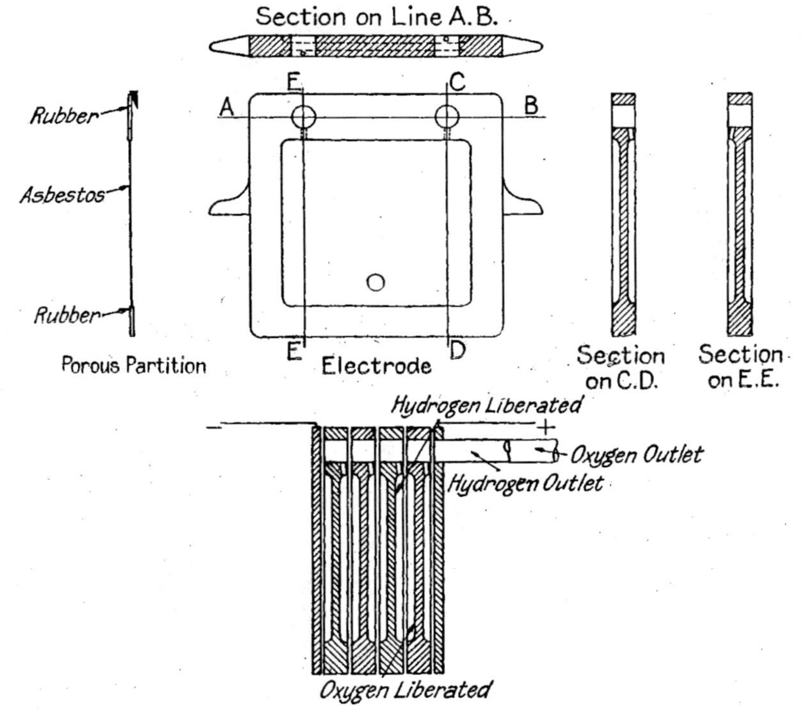

FIG. 15.

Filter Press Type.—If, in the diagrammatic cell (Fig. 15), a plate of conducting material was placed between the anode and cathode and the current switched on, hydrogen would be liberated at the original cathode and oxygen at the original anode, but, besides this, it would be found that on the side of the plate facing the original cathode oxygen would be liberated, while on its

FIG. 16.

other side hydrogen would be given off; thus it is seen that the intermediate plate becomes on one face an anode and on the other a cathode. Further, it will be found that the polarisation or back E.M.F. resistance of the cell from the original anode to the original cathode is doubled; thus the placing of a conductor, to which no electrical connections have been made, turns the original cell into two cells. The filter press cell is constructed on lines analogous to the above.

The filter press cell is composed of a series of iron plates, which are recessed on either side as shown in the diagram, from which it will be seen that, if two of these plates are put together, a space will be enclosed by them by virtue of the recess.

In each plate there are three holes, one at X and two along the line AB, so that, when the plates are placed together, the enclosed recess could be filled with water by means of the hole X. A small hole communicates with recess and the holes on AB, but in the case of one this communication is on the right-hand side while on the other it is on the left. Now, between any two plates is placed a partition, the shape and holes in which exactly coincide with those in the plates. The edge of this partition is composed of rubber, while the centre portion, which is of the same size as the recess in the plate, is made of asbestos cloth.

If four of these plates are pressed together with the partitions between, they will make three symmetrical cells which can be filled with electrolyte by blocking up the hole X in one outside plate and running it in through this hole in the other outside plate. Since the asbestos portion of the partition is porous, the electrolyte will soon reach the same level in each cell.

Now, if a positive electric connection is made to one outside plate and a negative to the other, what current passes must flow through the electrolyte and consequently electrolysis will take place. Since each plate is insulated from the other by the rubber edge of the partition each plate becomes on one face an anode and on the other a cathode, as was described in the diagrammatic cell, but the two plates which go to make the recess are divided by the asbestos partition, so the gases liberated have little opportunity of mixing. Since, as has been already mentioned, one of the holes in the top of the plate is in communication with one side of the recess and the other hole with the opposite side, the hydrogen and oxygen formed pass via separate passages to different gas-holders.

The description is applicable to all filter press type cells. The actual voltage of the electrical supply determines the number of plates which are in the complete unit, for the individual resistances are in series. In practice, using a 10 per cent. solution of caustic potash as the electrolyte, the voltage drop per plate is 2·3-2·5. The current density is generally about 18-25 amperes per square foot, while the production is 5·9 cubic feet of hydrogen and 3 cubic feet of oxygen, at mean temperature and pressure, per kilowatt-hour, the purity of the hydrogen being about 99·0 per cent. and that of the oxygen 97·5 per cent.[1]

The filter press type of cell has a considerable advantage by being compact, but, on the other hand, since

[1] The reason for the difference in purity is due to the fact that a small amount of diffusion takes place through the porous partitions, and since on account of its density the volume of hydrogen diffusing into the oxygen will be greater than the amount of oxygen diffusing into the hydrogen, the purity of the oxygen must of necessity be less than that of the hydrogen.

the water and gas tightness of the individual cells depends on the rubber in the partition and on the method of pressing the plates together, both these require a certain amount of attention ; probably a cell of this type would require overhauling in these particulars about once in every two and a half months, if it were kept running continuously.

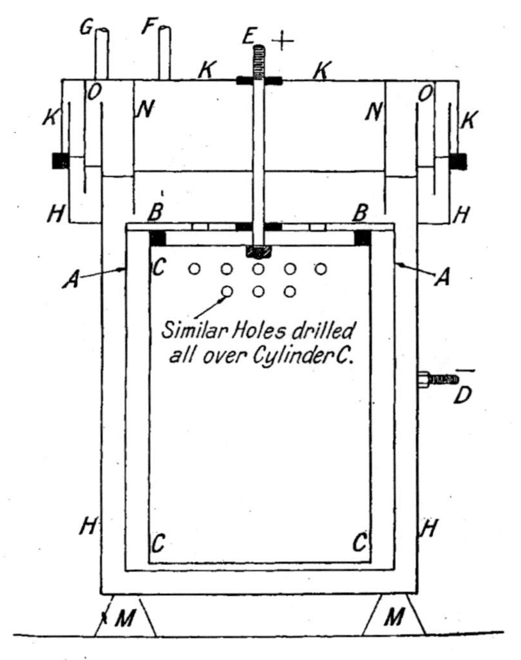

Fig. 17.—Tank Cell.

The following are probably the best-known commercial cells of this type : Oerlikon and Shriver.

The Tank Cell.—This type of cell will be readily understood by looking at the diagram (Fig. 17). It consists of a circular tank H made of dead mild steel, standing on insulators M, with an annular ring at the top. In this tank an iron cylinder C, perforated with holes, is hung from the cast-iron lid of the cell K by means of an electrode E. Between the side of the tank

FIG. 18.—International Oxygen Company's Cell.

H and the cylinder C an asbestos curtain A is hung from a plate of non-conducting material B. The lid of the tank, which is insulated from both H and C, has two flanges O and N which form an annular ring. It also has two outlet pipes G and F.

The annular space in the tank H is filled with water, while the interior of the tank is filled with a 10 per cent. solution of caustic soda in distilled water.

The method of operation of the cell is as follows: If the negative lead of the circuit is connected to D, which is metallically fastened to the tank body H, and the positive lead is connected to E, electrolysis will take place and hydrogen will be liberated on the side of the tank H, rising through the electrolyte into the annular space enclosed by the flanges N and O on the lid K, from whence it is free to circulate to the outlet pipe G. While hydrogen is being liberated on the sides of the tank H, oxygen will be liberated on both sides of the cylinder C, from whence it will rise up, ultimately finding its way through holes in the plate B into the annular space enclosed by the flange N, and thus on to the outlet pipe F.

There is a trapped inlet pipe (not shown) in the cover K for introducing further distilled water from time to time, to replace that decomposed by the operation of the process.

The voltage drop between anode and cathode is about 2·5 volts.

The outlet pipes G and F are usually trapped in a glass-sided vessel, which enables the working of the cell to be examined.

Fig. 18 shows a tank cell of the International Oxygen Company, which is not unlike the diagrammatic cell which has just been explained. Tests on four of these cells by the Electrical Testing Laboratories of New York give the following figures[1]:—

[1] Ellis, "The Hydrogenation of Oils" (Constable).

Average Amps.	Average Volts.	Average Watts.	Max. Temp. C.	Temp. 20° C. and Bar. 29·92 Ins.			
				Cubic Ft. per Hr.		Cubic Ft. per K.W. Hr.	
				Oxygen.	Hydrogen.	Oxygen.	Hydrogen.
392·7	2·609	1022	30·1°	3·114	6·075	3·051	5·950

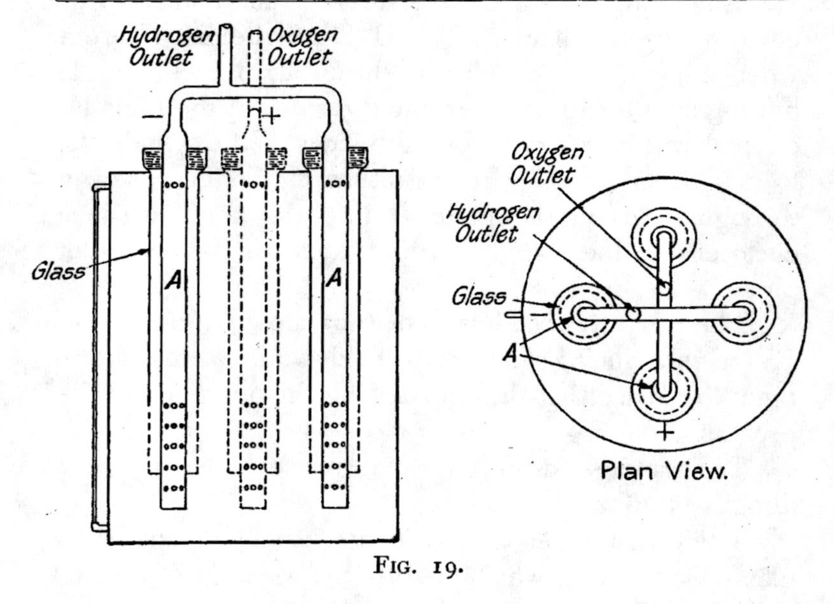

FIG. 19.

The purity of the oxygen was 98·34 per cent. and that of the hydrogen (from another test) 99·70 per cent.

The best-known plant of this type is that of the International Oxygen Company.

The Non-Conducting, Non-Porous Partition Type.—This cell, which is illustrated by the diagram (Fig. 19), consists of a metal tube A, which forms the electrode and gas outlet, and which is made of lead where an acid electrolyte is used, and of iron where an alkaline one is employed. This metal electrode is surrounded by a glass or porcelain tube perforated at the bottom.

There are four of these electrodes per cell, which are arranged as indicated in the diagram. When the current is switched on the gases are liberated on the electrodes within the glass tube ; consequently no mixing of the liberated gases can take place.

The best-known commercial cell of this type is the Schoop.

The Metal Partition Type.—In the preliminary description of the filter press type of cell it was stated that a conducting partition between the anode and cathode itself became on one face an anode and on the other face a cathode ; this, however, requires modification, as it is only true when the voltage drop between the original anode or cathode and the metal partition is less than the minimum voltage required for continuous electrolysis.

In the metal partition type of cell a metal partition is placed between the anode and cathode. This partition is insulated from the poles, is not so deep as the electrodes, and is perforated on the lower edge with small holes which, though reducing the electrical resistance, do not allow of the gases mixing.

The best-known cell of this type is the Garuti, which, since the true electrodes are only about half an inch apart, gives a more compact cell than if a non-conducting partition were employed.

Since the voltage drop between electrodes is slightly less for the same electrolyte than if a non-conducting partition were employed the yield is good, being about 6·1 cubic feet of hydrogen at mean temperature and pressure per kilowatt-hour. The current density is as high as 25 to 28 amperes per square foot, using a 10 per cent. solution of caustic soda.

The advantage of this type of cell is its compactness, due to the small distance between the electrodes, and its lightness, due to the fact that it is made throughout (with the exception of the insulating strips) of mild steel sheet. However, the small distance between the electrodes necessitates care being taken to prevent an internal short circuit in the individual cells.

Castner-Kellner Cell.—Besides those cells already described, the object of which is to produce oxygen and hydrogen, there are some which, though not designed for the production of hydrogen, yield it as a by-product.

Probably the most important of these electrolytic processes yielding hydrogen as a by-product is the Castner-Kellner. The primary purpose of this process is to make caustic soda from a solution of brine; but both hydrogen and chlorine are produced at the same time.

The working of this process can be understood from the diagram (Fig. 20).

The plant consists of a box A, divided into three compartments by the partitions B, which, however, do not touch the bottom of the box A. On the floor of this box there is a layer of mercury, which is of sufficient depth to make a fluid seal between the compartments. In the two end compartments there are carbon electrodes, connected to a positive electric supply, while in the middle there is an iron electrode, connected to the negative supply. One side of the box A is carried on a hinge H, while the other is slowly lifted up and down by an eccentric G, which gives a rocking motion to the contents of the box.

In the two end compartments is placed a strong solution of brine, while the middle is filled with water. On the current being switched on electrolysis takes place, the current passing from the positive carbon electrodes through the brine to the mercury, and from the

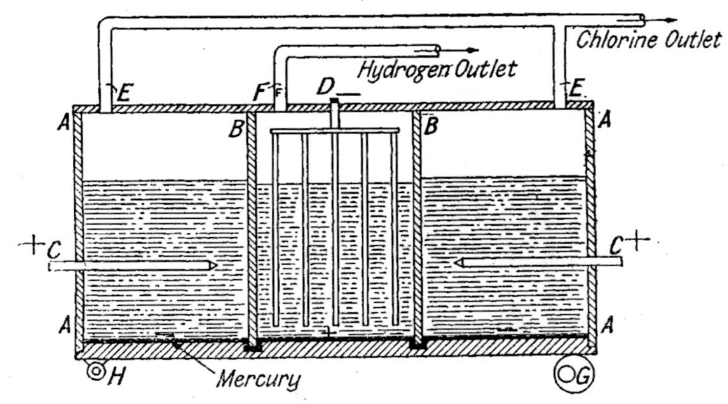

FIG. 20—Castner-Kellner Cell.

mercury to the negative electrode in the centre compartment.

Now, considering one of the end compartments, by the splitting up of the sodium chloride, chlorine will be liberated at the positive electrode, and will ultimately pass out at E, to be used for making bleaching powder, or for some other purpose, while sodium will be deposited on the mercury, with which it will amalgamate.

Owing to the rocking of the box, the sodium mercury amalgam will pass into the centre compartment, where it is decomposed at the negative electrode, in accordance with the following equation :—

$$2Na + 2H_2O = 2NaOH + H_2.$$

Thus, by the operation of the process, chlorine is produced in the end compartments, and caustic soda and hydrogen in the centre one.

The following patents have been taken out for the production of hydrogen electrolytically :—

Delmard.	German patent		58282.	1890.
Garuti.	English	,,	16588.	1892.
Baldo.	,,	,,	18406.	1895.
Garuti.	U.S.	,,	534259.	1895.
Garuti & Pompili.	English	,,	23663.	1896.
,, ,,	U.S.	,,	629070.	1899.
Schmidt.	German	,,	111131.	1899.
Hazard-Flamand.	U.S.	,,	646281.	1900.
Garuti & Pompili.	English	,,	12950.	1900.
,, ,,	,,	,,	2820.	1902.
,, ,,	,,	,,	27249.	1903.
Vareille.	French	,,	355652.	1905.
,,	U.S.	,,	823650.	1906.
Aigner.	German	,,	198626.	1906.
Cowper-Coles.	English	,,	14285.	1907.
Eycken Leroy & Moritz.	French	,,	397319.	1908.
Schuckert.	German	,,	231545.	1910.
Fischer, Luening & Collins.	U.S.	,,	1004249.	1911.
Moritz.	,,	,,	981102.	1911.
Hazard-Flamand.	,,	,,	1003456.	1911.
L'Oxhydrique Francaise.	French	,,	459957.	1912.

Benker.	French patent		461981.	1913.
Knowles Oxygen Co. & Grant.	English	,,	1812.	1913.
Maschinenfabrik Surth.	French	,,	462394.	1913.
Burdett.	U.S.	,,	1086804.	1914.
Ellis.	,,	,,	1087937.	1914.
,,	,,	,,	1092903.	1914.
Levin.	,,	,,	1094728.	1914.

APPENDIX.

PHYSICAL CONSTANTS.

Physical Properties of Hydrogen.

Critical temperature . . . . – 234° C.
,, pressure . . . . 20 atmospheres.
Melting point at atmospheric pressure – 259° C. } Travers, 1902.
Boiling point ,, ,, – 252·7° C. } Travers, 1902.

Density of Liquid Hydrogen.

At boiling point . . . . ·07
At melting point . . . . ·086

Vapour Pressure of Liquid Hydrogen (Travers & Jacquerod, 1902).

Temperature °C.	– 258·2	– 256·7	– 255·7	– 255·0	– 254·3	– 253·7	– 253·2	– 252·9
Pressure mm.	100	200	300	400	500	600	700	760

Latent Heat of Hydrogen.

123 cal. per grm.

DENSITY OF GASEOUS HYDROGEN.

At 0° C. and 760 mm.

·08987 grm. per litre.
5·607 lb. per 1000 cubic feet.

SPECIFIC HEAT OF GASEOUS HYDROGEN.

At constant pressure.

At atmospheric pressure . .	3·402	Lussana, 1894.
30 atmospheres . . .	3·788	

At constant volume.

At 50° C. . . . . . . 2·402 (Joly, 1891).

VELOCITY OF SOUND IN HYDROGEN.

At 0° C. = $12·86 \times 10^4$ cm. per sec. (Zoch, 1866).

SOLUBILITY OF HYDROGEN IN WATER.

The coefficient of absorption is that volume of gas (reduced to 0° and 760 mm.) which unit volume of a liquid will take up when the pressure of the gas at the surface of the liquid, independent of the vapour pressure of the liquid, is 760 mm.

Temperature. ° C.	Coefficient of Absorption.	Temperature. ° C.	Coefficient of Absorption.
0	·02148[1]	60	·0144[2]
10	·01955[1]	70	·0146[2]
20	·01819[1]	80	·0149[2]
30	·01699[1]	90	·0155[2]
40	·0152[2]	100	·0166[2]
50	·0146[2]		

TRANSPIRATION OF GASEOUS HYDROGEN.

Oxygen	1·0
Hydrogen	·44

[1] Winckler (Ber., 1891, 99).
[2] Bohr and Beck (Wied, Ann., 1891, 44, 316),

REFRACTIVITY OF HYDROGEN.

	$\mu - 1$	
Air	1·000	Ramsay & Travers.
Hydrogen	·473	

RELATIONSHIP BETWEEN PRESSURE AND VOLUME.

Were Boyle's Law correct then the product of the pressure multiplied by the volume would be a constant; however, Boyle's Law is only an approximation, all gases near to their critical temperature being much more compressible than the law indicates. At atmospheric temperature the common gases, such as oxygen and nitrogen, are very slightly more compressible than would be expected from theory. Hydrogen and helium under the same conditions are less compressible, hence Regnault's description of hydrogen as "gas plus que parfait".

The behaviour of hydrogen at low pressures (from 650 to 25 mm. of mercury) was investigated by Sir William Ramsay and Mr. E. C. C. Baly, who found that, at atmospheric temperature, Boyle's Law held throughout this range of pressure.

The relationship between volume and pressure when the latter is great has been investigated by Amagat and Witkowski, whose results are incorporated in the graph (Fig. 21), which shows the relationship between the theoretical volume of hydrogen which should be obtained on expansion to atmospheric pressure and that which is obtained from a standard hydrogen cylinder. From this it is seen that, on expansion from 2000 lb. per square inch to atmospheric pressure, 9·2 per cent. less volume of hydrogen is obtained than is indicated by theory.

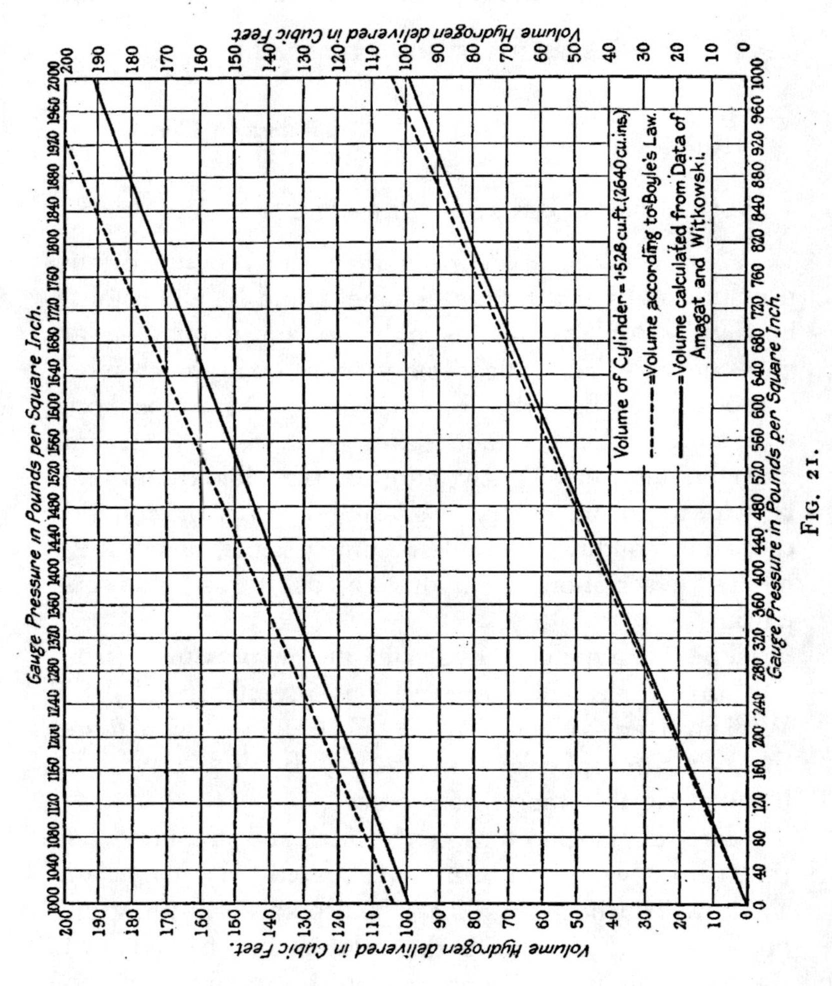

FIG. 21.

THE JOULE-THOMSON EFFECT.

Down to at least − 80° C. hydrogen on expansion by simple outflow rises in temperature, which is unlike all other gases with the possible exception of helium. The variation in temperature for drop in pressure of unit atmosphere for air and hydrogen is given below :—

	T. °C.	[1] Variation per Atmospheric Pressure.
Air	17·1	− 0·255° C.
	91·6	− 0·203
Hydrogen . .	6·8	+ 0·089
	90·3	+ 0·046

Lift of Hydrogen.—Lift of 1000 cubic feet of hydrogen $= \dfrac{12\cdot34 \times P \times B}{460 + T}$ lb.

where P = Purity of hydrogen by volume expressed in percentage.
B = Barometric pressure in inches.
T = Temperature of air in degrees Fahrenheit on the dry thermometer.

This formula is correct if the air is dry. If it is wet a small correction must be applied, which is given in the following curve.

The purity of the hydrogen is expressed by volume on the assumption that the impurity is air or some other gas of the same specific gravity as air under the same conditions; if the impurity is not air due allowance must be made.

Correction for Humidity of Air.—The attached curve gives the correction which must be employed in the lift formula for humidity of the atmosphere. The difference between the temperature of the air on the wet and dry thermometers is found on the left-hand side of the graph; the temperature of the air as shown on the dry thermometer is found on the bottom; find where perpendiculars from these two points intersect and

[1] Joule and Lord Kelvin.

estimate the value of the correction from the position of the point of intersection relative to the curved lines.

EXAMPLE.—Let the air temperature be

Dry.	Wet.
60° F.	50° F.

then difference is 10° F., and the intersection of the perpendiculars is between the curved lines ·35 and ·4 at

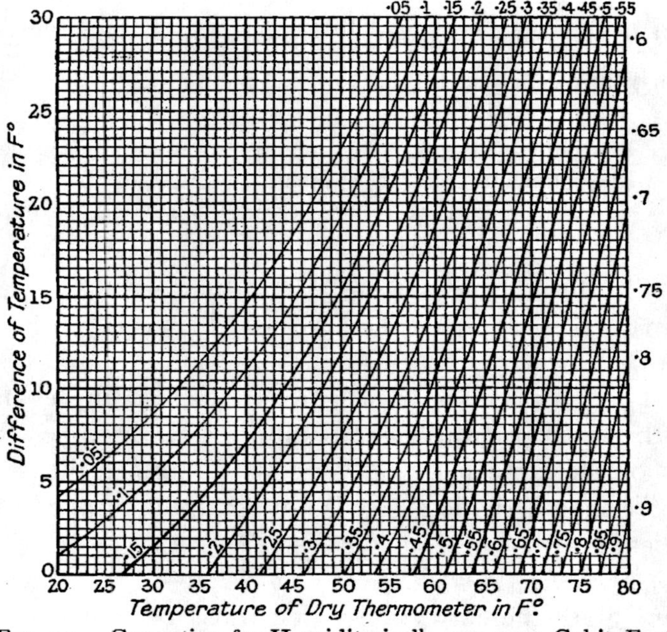

FIG. 22.—Correction for Humidity in lb. per 1000 Cubic Feet.

a position which may be estimated at ·36 lb.; therefore 36 lb. must be subtracted from the lift per 1000 cubic feet of hydrogen as determined by the formula when the temperature of the air by the dry thermometer was 60° F. and the difference between wet and dry 10° F.

INDEX.

ABSORPTION of hydrogen by metals, 15.
Air, composition of, 7.
— hydrogen in, 7.
Aluminal process, 44.
Ammonia, 26.
— absorption by charcoal, 29.
— liquefaction, 29.
— properties, 27.
— solubility, 28.
— uses, 27.
Arsine, 32.
— production in Silicol process, 32.

BADISCHE Catalytic process, 101.
— — — patents, 106.
— — — plant, 105.
— — — preparation of catalyst, 103.
Bergius process, 63.
— — patents, 66.
Boiling point of gases, 115.
— — — hydrogen, 145.
Bronze, hydrogen in, 4.

CALCIUM hydride, 34.
Carbonium-Gesellschaft process, 108.
Centrifugal separation of hydrogen, 124.
Cerium hydride, 34.
Clays, hydrogen in, 7.
Critical pressure, 114.
— — of hydrogen, 145.
— temperature, 114.
— — of hydrogen, 145.

DENSITY of gaseous hydrogen, 145.
— — liquid hydrogen, 145.
Diffusion, separation of hydrogen by, 121.
Discovery of hydrogen, 2.
Draper effect, 22.

ELECTROLYTIC cells—
Castner-Kellner cell, 142.
filter press type, 133.
metal partition type, 141.
non-conducting, non-porous partition type, 140.
patents, 144.
tank cell, 137.
Electrolysis, 126.
Explosions of mixtures of hydrogen and oxygen, 14.

FAT hardening, 35.
Ferro-silicon, 50.

HEAT produced by ignition of hydrogen and oxygen, 17.
Hydrik process, 44.
Hydriodic acid, 24.
Hydrobromic acid, 23.
Hydrochloric acid, 21.
Hydrogen and arsenic, 32.
— — bromine, 23.
— — carbon, 20.
— — chlorine, 21.
— — iodine, 24.
— — nitrogen, 26.
— — oxygen, 9.
— — phosphorus, 30.
— — selenium, 25.
— — sulphur, 24.
— — tellurium, 26.
— physical constants, 145.
— production. *See* Production of hydrogen.
Hydrogenite process, 60.
Hydrolith process, 67.

IGNITION temperature of hydrogen and oxygen, 10.
Iron Contact process, 86.
Fuel consumption, 97.
Oxidising, 90.
Patents, 99.
Plant—
Multi-retort type, 93.
Single retort type, 95.
Purging, 89.
Purification of hydrogen, 90.
Reducing, 88.
Secondary reactions, 91.
Sulphuretted hydrogen in, 93.

JOULE-Thomson effect, 148.

LATENT heat of hydrogen, 145.

Lift of hydrogen, 149.
Linde-Frank-Caro process, 113.
— patents, 121.
— purification of gas, 120.
Lithium hydride, 33.

MAGNESIUM hydride, 34.
Manufacturing processes—
Badische Catalytic, 101.
Bergius, 63.
Carbonium-Gesellschaft, 108.
Electrolytic, 132.
Hydrik, 44.
Hydrogenite, 60.
Hydrolith, 67.
Iron Contact, 86.
Linde-Frank-Caro, 113.
Sical, 69.
Silicol, 45.
Meteoric iron, hydrogen in, 3.

OCCURRENCE of hydrogen, 2.
Oil and gas wells, hydrogen in discharge from, 5.
Oxygen, explosion of hydrogen and, 14.
— heat produced by ignition of hydrogen and, 17.
— ignition temperature of hydrogen and, 10, 15.
— reaction of hydrogen with, 9.

PHOSPHINE, 30.
— action on metals, 31.
Phosphoretted hydrogen, 30.
Physical constants of hydrogen, 145.
Polarisation resistance, 130.
Potassium hydride, 33.
Production of hydrogen, 39.
— from acetyline, 108.
— — acid and iron, 40.
— — acid and zinc, 42.
— — alkali and aluminium, 44.
— — — — carbon, 60.
— — — — formate, 60.
— — — — oxalate, 61.
— — — — silicon, 45.
— — — — zinc, 43.
— — water and aluminium alloy, 71.
Production from water and aluminium amalgam, 69.
— — — — — — silicide, 68.
— — — — metallic hydrides, 66.
— — — — metals, 61.
— — hydrocarbon oils, 110.
— — starch, 111.
— — steam and barium sulphide, 100.
— — — — iron, 86.
— — — — water gas, 101.

REFRACTIVITY of hydrogen, 147.
Rocks, hydrogen in, 3.

SELENURETTED hydrogen, 25.
Sical process, 69.
Silicol process, 45.
composition of sludge, 55.
lime, use of, 53.
mineral grease, use of, 57.
patents, 59.
plant, 47.
precautions to be taken, 57.
purity of hydrogen produced, 45.
strength of caustic, 52.
Sodium hydride, 33.
Solubility of hydrogen in water, 146.
Sound, velocity in hydrogen, 146.
Specific heat of hydrogen, 146.
Sulphuretted hydrogen, 24.
— — removal from water gas, 84

TELLURETTED hydrogen, 26.
Transpiration of hydrogen, 146.

USES of hydrogen, 1.

VOLCANOES, hydrogen in gases from, 5.

WATER gas manufacture, 72.
— — — Dellwick method, 75.
— — — English method, 74.
— — — Swedish method, 75.
— — purification of, 82.
— — removal of sulphuretted hydrogen from, 84.